Ways of Knowing in Science Series

RICHARD DUSCHL, SERIES EDITOR

ADVISORY BOARD: Charles W. Anderson, Nancy Brickhouse, Rosalind Driver,
Eleanor Duckworth, Peter Fensham, William Kyle,
Roy Pea, Edward Silver, Russell Yeany

Inventing Science Education for the New Millennium
Paul DeHart Hurd

Improving Teaching and Learning in Science and Mathematics
David F. Treagust, Reinders Duit, and Barry J. Fraser, Editors

Reforming Mathematics Education in America's Cities:
The Urban Mathematics Collaborative Project
Norman L. Webb and Thomas A. Romberg, Editors

What Children Bring to Light:
A Constructivist Perspective on Children's Learning in Science
Bonnie L. Shapiro

STS Education: International Perspectives on Reform
Joan Solomon and Glen Aikenhead, Editors

Reforming Science Education:
Social Perspectives and Personal Reflections
Rodger W. Bybee

Inventing Science Education

for the

NEW MILLENNIUM

Paul DeHart Hurd

Teachers College, Columbia University
New York and London

For E. K. H., whose encouragement and
help made the writing of this book possible

Published by Teachers College Press, 1234 Amsterdam Avenue, New York, NY 10027

Portions of Chapter 3 were abstracted from "Closing the Educational Gaps Between Science, Technology, and Society." *Theory into Practice.* 30 (4), 251–259 by Paul Hurd. Copyright © 1991 College of Education, the Ohio State University.

Chapter 5 is slightly modified from the article of the same title published in *Bulletin of Science, Technology, Society*, 14 (3), 125–131 by Paul Hurd. Copyright © 1994 STS Press.

Library of Congress Cataloging-in-Publication Data
Hurd, Paul DeHart, 1905–
    Inventing science education for the new millennium / Paul DeHart
Hurd.
        p.   cm. — (Ways of knowing in science series)
    Includes bibliographical references (p.) and index.
    ISBN 0-8077-3672-4 (cloth). — ISBN 0-8077-3671-6 (paper)
    1. Science—Study and teaching—United States.   2. Science—Study
and teaching—United States—Planning.   3. Educational change—
United States.   I. Title.   II. Series.
LB1585.3.H87   1997
507.1'073—dc21   97-23226

ISBN 0-8077-3671-6 (paper)
ISBN 0-8077-3672-4 (cloth)

Printed on acid-free paper
Manufactured in the United States of America

04   03   02   01   00   99   98   97   8   7   6   5   4   3   2   1

Contents

Preface

This book is offered as the beginning of a discourse on the teaching of science for life in a new millennium. A description of the major changes taking place in the ethos and culture of science as well as in the nation's social structure are identified. These attributes serve as a model of the contexts essential for inventing science curricula in harmony with contemporary science, our changing culture, and our quality of life.

ACKNOWLEDGMENTS. Over the past five years I have interviewed over 150 junior and senior high school teachers about their notions on the current science education movement. For the most part they saw the movement as simply the bashing of schools and teachers for no good reason. It was these comments that set the agenda for this book.

I am grateful to Lori Tate and the production staff of Teachers College Press for seeing that my writing expresses the intended idea.

CHAPTER 1

Prologue

Since midcentury it has been evident that in various ways our nation and most other countries are entering a new era of existence. The roots of this transition lie in achievements of contemporary science and technology. These developments are changing the character of our culture and how people live, learn, and work. These factors have brought demands on schools to do something about the curriculum and instruction in the sciences to make it more in harmony with the changes taking place in our society and in the nature of science.

In the late 1950s the National Science Foundation (NSF) took on the task of improving school science courses at all grade levels. Two or more programs were developed for each grade or subject. The goals and an outline of the subject matter for each course were published by Hurd and Gallagher, 1968, and Hurd, 1969a,b and 1970a,b. The projects focused on updating the subject matter in the various disciplines represented in school science courses. The developers emphasized that students should learn inquiry skills characteristic of those used by research scientists. To improve laboratory skills, they also saw a need for exercises that had elements of research in the sciences and that would develop skills for using modern laboratory equipment. Today these courses are viewed as discipline bound and occupationally oriented. They serve science more than they serve people and society.

In the 1970s scholars in a variety of academic fields recognized that the United States was experiencing massive critical changes in nearly all aspects of our lives—socially, culturally, and economically, and especially in the revolution taking place in science and technology. The knowledge and the intellectual skills students will use to function effectively in the years ahead were not found in elementary and secondary school science courses.

Not everyone saw these changes in the same way, but all recognized that we are experiencing an age not like any of the past. A story line reflecting the intellectual mood of scholars regarding our changing society can be found in the writings of the following authors. Robert Hutchins (1968), president of the University of Chicago, noted that the nation was moving toward "a learning society"; Zbigniew Brzezinski (1970), of the U.S. State Department, perceived the nation as "between two ages"; Peter Drucker (1968), economist, New York University, described the social changes taking place in our society as representing a "discontinuity" in the nation's history; Kenneth E. Boulding (1964), an economist and former president of the American Association for the Advancement of Science (AAAS), characterized the new era as a "cultural mutation"; and Donald N. Michael (1968), of the Center for Research on the Utilization of Scientific Knowledge at the University of Michigan, defined our society as one for which we were "unprepared." Dennis Gabor (1964), a Nobel laureate in physics, saw the need for this new society "to invent its future."

Seeing what happens to people when they are overwhelmed with changes taking place in their lives led Alvin Toffler (1970) to write his book *Future Shock*, concerning the need to identify new directions for living, society, and education. Charles E. Silberman (1970) viewed the influences of these changes as creating "a crisis in the classroom." These authors and other scholars have warned us that the traditional interpretations of our culture, of society, of science and technology, and of our lifestyles and how people live, work, learn, and communicate are different from what they were at any time in our past. Each of these factors has implications for creating a new vision of science education, one that serves to connect students to a changing culture.

The task of a reformation in science education is made difficult by revolutionary changes in the nature and practice of the traditional sciences over the past half-century that are continuing at a rapid rate. At the same time, hundreds of new sciences have been created that are unrepresented in school science curricula. Noteworthy is the number of new sciences that focus on human welfare, and on social and economic progress. Paralleling the changes in science and society are those associated with family lifestyles and their impact on childhood and adolescence. The changes have created new demands on the education children and adolescents must have.

Altogether these changes create the framework for a reformation of education in the sciences. The curriculum reform efforts sponsored by the National Science Foundation during the 1950s and 1960s have proved not to be rich enough in concept and subject matter to meet the demands of living in the twenty-first century. In 1970 the Advisory Committee for Science Education of NSF, recognizing the changes taking place in science and society, reversed its previous position on the purpose of science education as occupational training for the preparation of scientists to one that states, "To educate scientists who will be at home in society and to educate a society that will be at home with science" (NSF Advisory Committee for Science Education, 1970).

Where do we go from here? Insights into the problems and issues we must face to bring about valid changes are the theme of this book. This prologue is intended to provide some understanding of the issues we confront in transforming education in the sciences. Our task is one of long standing. Aristotle aptly described the primary issue we must deal with today. He wrote these notes after surveying the need for educational reform in the schools of Athens in 300 b.c. (quoted in Jewett & Butcher, 1964).

> That education should be regulated by law and should be an affair of state is not to be denied, but what should be the character of this public education, and how young persons should be educated, are questions which remain to be considered. As things are, there is disagreement about the subjects. For mankind are by no means agreed about the things to be taught, whether we look to virtue or the best life. Neither is it clear whether education is more concerned with intellectual or with moral virtue. The existing practice is perplexing; no one knows on what principle we should proceed—should the useful life, or should virtue, or should the higher knowledge, be the aim of our training; all three opinions have been entertained. (Jewett & Butcher, 1964, p. 268)

The American public has made recommendations in over 400 national reports, a host of articles in newspapers and weekly magazines, and television programs, and in other ways. A normative analysis of these reports has yet to be completed. A variety of science/technology agencies, around 80, supported by the national government have been asked to focus some attention on the reform of science education. As we approach the twenty-first century, we find the science education movement taking place along two fronts, (a) the ideas proposed by most educators and the public, and (b) proposals being formulated by the U.S. government. The public and the government do not see the reform issues in the same way. Both sets of views are presented in Chapter 2 of this book.

To gain a better perspective on the demands for a science education reform, it is helpful to explore briefly its roots. This is the subject of Chapter 3.

Changes in the nature of contemporary science/technology are portrayed in Chapters 4, 5, and 6.

A much neglected factor in the reform movement so far has been the student. Throughout the twentieth century, adolescents have become increasingly roleless in our society. The sciences that could help them understand themselves and improve their adaptive capacities are rarely noted in the reform literature. Yet there has been extensive study and research on understanding today's adolescents. These findings are presented in Chapter 7, along with implications for new curricula.

Science teaching for a changing world is outlined in Chapter 8. Here the changing concepts of what learning means in terms of knowing, understanding, and the utilization of knowledge are considered.

Chapter 9, the epilogue, deals with problems associated with the implementation of new science curricula for the twenty-first century. It seems apparent that it will be necessary to develop a broader leadership for science education. A plea will be made for the national government to establish and support a national academy of science education on a permanent basis.

It is not my intention in this book to criticize science teaching and curricula now found in schools. These curricula, their goals, and the supporting teaching practices represent a period of our nation's educational history that can no longer be justified due to changes in our society and in the sciences. My purpose is to open the doors to the impact of contemporary science and the forces of technology that are shaping our culture and creating a demand for a new vision of science education. The emphasis throughout the book is on changes that should influence precollege education in the sciences.

Physicist J. Robert Oppenheimer described the nature of these changes in a recorded talk, "Prospects in the Arts and Sciences" at a University of Colorado summer session.

> In an important sense this world of ours is a new world, in which the unity of knowledge, the nature of human communities, the order of society, the order of ideas, the very notions of society and culture have changed and will not return to what they have been in the past. What is new is new not because it has never been there before, but because it has changed in quality. One thing that is new is the prevalence of newness, the changing scale and scope of change itself, so that the world alters as we walk in it, so that the years of man's life measure not some small growth or rearrangement or moderation of what he learned in childhood, but a great upheaval. What is new is that in one generation our knowledge of the natural world engulfs, upsets, and complements all knowledge of the natural world before. (Oppenheimer, 1963)

Albert Einstein is reported as having stated that the significant problems we face cannot be solved at the level of thinking we were at when we created them.

The recurring themes of this book are the changes taking place in our culture, in the practice of science, and in the lives of students. These changes in one way or another interact and call for a rethinking of the purposes of an education in the sciences for an American renaissance.

CHAPTER 2

Science Education Reform on Two Fronts

Efforts to reform science education have now been under way for a quarter of a century. The problem is seen as extending from preschool through doctoral programs in universities. The movement is worldwide. UNESCO has identified 141 countries now engaged in rethinking educational programs in the sciences. This problem emerges from the wealth of social, economic, and conceptual changes in the nature of science that require new educational policies.

The appropriate mood for reading the rest of this chapter was stated by Niccolò Machiavelli (1513/1977): "There is nothing more difficult to manage, or more dangerous to execute, than the introduction of a new order of things" (p. 75).

Over the past 50 years, science—and consequently, science education—have become imbedded in a maze of contexts. These contexts include social, cultural, political, economic, technological, behavioral, historical, cognitive, and evolutionary changes in the nature of science, as well as in the implications of science as it relates to living. During this half-century, research in the natural sciences has become more socially driven than theory driven. The mosaic of changes taking place in the nature and practice of contemporary sciences is making obsolete the current goals and curricula of school science.

The growing dissatisfaction with school science exploded in the early 1970s. Of particular concern today is the realization that young people are leaving school ill prepared for life in the modern world in which they must spend their lives. Students are seen as lacking the intellectual skills for coping and adapting to our changing culture. Glenn Seaborg, a Nobelist in chemistry, has commented in several talks that students are being graduated from high school today as "foreigners in their own culture."

School science curricula as they now exist are perceived as too narrow in vision goals and subject matter to have much meaning beyond the laboratory door. The traditional practice of teaching school science in the context of a career in science, a way of developing amateur scientists and preparing students for college, is being replaced by a concept of the public understanding of science. This approach reduces the traditional emphasis on "thinking like a scientist," "acting like a scientist," acquiring "science process skills," learning about "scientific inquiry," and "preparing for college." It should be noted that these procedures are not part of how career scientists are educated today (NAS, 1995).

Many of the concerns now being raised about science education have been fermenting for most of the twentieth century. The current reform is distinguished by the extent of public awareness as well as by national pressures for a new vision of public education in the United States. During the 1980s there were close to 400 national reports recommending educational improvements. These reports were prepared by private and public educational foundations, government agencies, business and economic organizations, parent associations, and special interest groups. In most of these reports a central issue was education in the sciences. What was lacking was a cohesive view of what this education should comprise.

In 1970 the Advisory Committee for Science Education (ACSE) of the National Science Foundation reversed its position on science education from that of precollege discipline-based courses to interdisciplinary society-oriented courses. The report stressed one overriding goal: "To educate scientists who will be at home in society and to educate a society that will be at home with science." Also, courses should be "centered upon problems faced by informed citizens" (NSF Advisory Committee for Science Education, 1970).

In 1996 Neal Lane, director of the National Science Foundation, raised the question "The public likes science, but do scientists like the public?" (Lane, 1996a). He later editorialized in *Science* that new "leadership is needed from those of us in the research community—particularly from individual scientists and engineers active in research—to carry our understanding of science and its value into the lives of all Americans." In this new environment of ours without an understanding of contemporary science and technology among citizens and policy makers, "science and the American dream may be only a memory of the past" (Lane, 1996b).

Task forces seeking to bring about a reformation of education in the sciences have proceeded along two contrasting fronts. A policy group sees the need to develop a new conceptual framework for thinking about the purposes of an education in the sciences for all citizens. Most of these deliberations are reflected in reports generated at the national level. Their emphasis is mostly centered on educational policies that reflect our changing culture and the nature and practice of today's science and technology.

Another group of reformers, the majority, see the quest for improving science education as one of dealing with problems that beset schooling and the teaching of science. Typical concerns are the need to update subject matter, making courses more rigorous, getting back to basics, modernizing equipment, especially encouraging the use of computers, reducing class sizes, and raising test scores on achievement tests both national and international. These problems are viewed at the local and state levels (Doyle & Hartle, 1985).

The public and the U.S. government have identified their concerns about the reform of education. Professional science educators and classroom teachers have been less clear about their perspectives for the reform of science education, especially in terms of a knowledge-intensive society and a changing culture. For the most part, they have chosen to recycle traditional goals and practices with little more than a "new paint job."

VIEWS REPRESENTATIVE OF POLICY GROUPS

Policy groups seek to examine the patterns and web of events that citizens face in this world of changes, particularly those brought about by revolutionary changes in science and technology. The report commanding the widest attention was *A Nation at Risk* (Gardner, 1983). Here the central purpose of education was seen as one in which "all children, by virtue of their own efforts competently guided, can hope to attain the mature and informed judgment needed to secure gainful employment and to manage their own lives, thereby serving not only their interests but also the progress of society itself" (Gardner, 1983). In the same year, the National Science Foundation published a review of the "scope, level, and direction of U.S. scientific and technological efforts." The focus of the report is on the generic theme of economics as a basis for the roles of science and technology in contemporary American society (NSF, 1983b).

In 1991 the U.S. government targeted precollege education in the sciences as a priority in the reform of all education, kindergarten through university. To accomplish the desired results, a special group, the Committee on Education and Human Resources, was established as a branch of the Federal Coordination Council for Science, Engineering, and Technology (FCCSET). Its purpose is to prepare "a comprehensive baseline inventory of federal funding and programs that affect mathematics and science education at all levels—precollege, undergraduate, and graduate." At the same time, FCCSET is to develop a set "of strategic objectives with budget priorities to guide future federal actions in this critical area" (FCCSET, 1991). The work of this committee involves coordinating 19 departments and offices of the national government which have programs related to some aspect of science and technology and the development of human capital, such as the Department of Health and Human Services, the National Science Foundation, the Smithsonian Institution, the Departments of Labor, Energy, and Agriculture, and the Office of Science and Technology Policy.

The movement of science education reform to a national government position has had political support from the beginning. It was Bill Clinton, as governor of Arkansas, who wrote the national goals for education while attending the education summit called by President George Bush in 1990. In 1991 Congress created the National Council on Education Standards and Testing (NCEST, 1992). The council began its work by defining content standards as broad goal statements that describe the knowledge, skills, and other understandings that schools should teach in order for students to attain high levels of competency in challenging subject matter (NCEST, 1992). A major skill recommended was "learning to learn."

In 1993 the U.S. Senate passed H. R. 1804, officially authorizing and financially supporting the tenets of the National Educational Goals Panel (H.R. 1804, 1993). The details of H. R. 1804 are elaborated and clarified in the Goals 2000: Educate America Act (1994). In July 1994 President Clinton signed this act into law.

As a reference for members of the U.S. House of Representatives in debating educational issues, a 300-page report was prepared summarizing most of the national reports on educational reform published during the 1980s, *Education 2005: The Role of Research and Development in an Overwhelming Campaign for Education in America* (Ford, 1991). It proposed a plan for educational improvement consisting of five years of widespread comprehensive experimentation; five years of intensive evaluation, modification, and certification; and five years of maximizing dissemination, consolidation, and institutionalization of "what works." If they achieved what they had set out to do, a new era for American education could then commence in the year 2005.

Bill Clinton and Al Gore have publicly supported educational reform from before President Bush's education summit. Their policy concerns include the importance of linking education to work as essential for personal and social economic opportunity; of parent participation in the education of children; of students' being knowledgeable in math and science, or acquiring decision making skills; bringing business, labor, and education leaders together in developing work skills; and of students "learning to learn." These views are expressed in their book *Putting People First: How We Can All Change America* (Clinton & Gore, 1992).

Their second book, *Technology for America's Economic Growth: A New Direction to Build Economic Strength,* focuses on the importance of technology and science for America's economic future, quality of life, and education and training (1993). The third book in their series, *Science in the National Interest* (1994), sets a national policy for the place of science and technology in education. The authors link precollege science education with research in the sciences and technology, and both science education and research with the nation's economic prosperity and work and a wide variety of human affairs. In a personal letter to the author (October 21, 1994) Gore summarized the position of the White House on the reform of science education:

> Our report proposes a series of actions to meet five broad goals for world leadership in science, mathematics, and engineering. First, we must maintain U.S. leadership across the frontiers of scientific knowledge. Second, we must work to enhance connections between fundamental research and national goals. Third, it is essential that we stimulate government, industry and academic partnerships that promote investment and fundamental research and national goals. Fourth, it is essential that we stimulate government, industry, and academic partnerships that promote investment in fundamental science and engineering and effective use of human, and financial resources. Fifth, we are determined to produce the finest scientists and engineers for the twenty-first century. Finally, we must raise the scientific and technology literacy of all Americans.

The task for educational reform in the sciences is generally seen as one of developing a vision recognizing that science, technology, society, and the quality of human existence are interconnected, and that their traditional borders are disappearing.

Increasingly, national reports emphasize the need to relate science education to the changing world of work that young people now face. Traditional school science curricula show career concern only for those students planning to become scientists. This has been true throughout the history of science education in the United States.

Today, more than anything else, it is the nation's movement from an industrial smokestack economy to a postindustrial knowledge-intensive society that influences the purposes of science education. The traditional brawn, mechanical skills, and natural resources needed to survive economically in our current knowledge-intensive age have little value. A major issue in reforming science education for the year 2000 is how to connect students with today's world of work. What is required is an assortment of higher level thinking skills a number of which can be developed from a study of science in a work context. These are also skills required of students to achieve economic success in life. National agencies speak of this goal as building human potential and talent, developing human capital, or increasing human resources, all in terms of more productive individuals and an advancing society (Hurd, 1989b).

Historically there has been a separation between schooling and work, although working is what people do most of their lives. Closing the gaps by linking school and work has become the most sought after national goal in the reform of science education (Jennings, 1995; Resnick & Wirt, 1996). It was evident from the beginning of the science education reform movement that our changing culture would require people to learn, think, and work differently from the past. These factors have made "learning to learn" an international goal for replacing the traditional concept of vocational education.

In 1990 the U. S. Department of Energy formed the Secretary's Commission on Achieving Necessary Skills (scans) to examine the workplace and to determine generic workplace competencies and how they might be achieved. The commission has issued a series of reports on the problem (scans, 1991, 1993). The theme of these reports is that in a knowledge-intensive society the resulting transformation of work becomes a matter of "learning a living." An outstanding book by Marshall and Tucker, *Thinking for a Living* (1992), amplifies the tensions taking place in our democracy and their influence on generating education qualities for working. The National Research Council (nrc) has explored current and needed research on learning to think (Resnick, 1987). My favorite bumper sticker comment is "If you feel college is expensive, compare it with the cost of ignorance."

Hurd (1989b) analyzed 50 reports developed by business, national commissions, and government agencies on educational changes needed to meet the challenges of our changing economy and the requisites for working. Of the 50 reports, only 26 specified work goals for every student; all the reports recognized that high school graduates today are ill prepared for work in today's economy. What are seen as lacking are students with higher order thinking skills and science curricula with modern social, economic, and technology dimensions. Problems of linking school and work have been explored by Resnick and Wirt (1996).

Another dimension of the science education reform movement is its role in augmenting an understanding of a science/technology–oriented democracy. The goal is a public understanding of science as the nation moves to a new era in our culture. These changes, appropriately directed, assure social and economic progress and a higher quality of life. In 1977, the National

Academy of Sciences (NAS) published a summary of commentaries from Academy meetings expressing views on the place of science in influencing national affairs (NAS, 1977). In general it was noted that research in the sciences should be more socially directed. A similar study by the Academy in 1993 also sought to explore science policy issues.

The American Association for the Advancement of Science (AAAS) viewed the national movement for the reform of science education as an effort to focus on ways of adapting science/technology to social needs (Scribner & Chalk, 1977). At the same time it is recognized that concepts of our democracy must be restudied to include science and technology as major factors in determining public policy and citizens' understanding of their culture and their responsibilities. The report notes that national objectives for research in the sciences should be determined by "criteria external to the field of research," such as criteria related to health, the environment, public welfare, and social and economic progress.

The Carnegie Commission on Science, Technology, and Government was formed to assess processes by which the U.S. government could incorporate science and technology into policy and decision making in the affairs of the nation (Carnegie, 1988). What this means for the reform of K–12 math and science education is described in a 1991 Carnegie report. A study by the Carnegie Commission focuses on linking science and technology to societal goals (Carnegie, 1992a). A further study addressed the role of states in fostering changes and the implementation of science and technology policies—including education—in transforming the effects of science and technology in our society (Carnegie, 1992b). An additional report portrays the roles of the President and Congress in bringing about a new level of participation for restructuring education in ways that safeguard the environment, making industry more competitive and ensuring national security (Carnegie, 1992c). One Carnegie study (1993b) emphasizes that:

> Science is not a separate entity from the lives of people. Indeed, science provides the basis for most of the requirements for modern living. The world has been transformed by science and technology in this century and this transformation is continuing, even accelerating, as the century comes to its close. (p. 4)

This view is reinforced by a report from the National Academy of Sciences (NAS, 1993).

A summary of the foregoing reports views science and technology as playing a broader role in the future in the social and institutional systems of which they are a part. This is a position seldom reflected in school science textbooks. Currently, most research in the sciences is focused on problems beyond a discipline, such as public welfare, social and economic progress, health, and the environment. National efforts along these lines have been proposed in Congress and in citizen groups. The National Science Foundation in a series of six reports (RANN [Research Applied to National Needs] 2, 1976) explored these issues in relating science and technology to social programs and for directing science research in the same vein.

What became evident was the importance of science/technology for building human capital in the United States, which has become a new goal for school science instruction. In turn, this means that the current career objectives of school science as they relate to the training of amateur scientists are too limited for our emerging culture. More significant are attitudes and skills which can be taught in science courses that are relevant for learning, living, and working in our knowledge-intensive society. What is now meant by career education for everyone is encompassed in the phrase "learning a living" (Sherman, 1983). The U.S. Department of Labor, in its reports on learning a living, has identified a range of intellectual and social skills and tasks essential for one to become a productive person in today's workforce (H.R. 2884, 1994).

The U.S. Department of Energy and the Federal Coordinating Council for Science, Engineering, and Technology (FCCSET), representing 18 divisions of the federal government, in 1991 proposed "a partnership program representing federal, state, and local governments; educators and parents; business and industry; professional associations; and community-based organizations." The program, *By the Year 2000: First in the World*, has as its overall goal to build a "nation of learners" and strengthen human resources in science, mathematics, and engineering. Every year from 1991 to the year 2000 there was to be an annual report identifying problems, issues, and achievements in the program. One of the first problems the committee identified was that science teachers are rarely able to relate scientific concepts to real-life or real-world situations.

In 1992 the National Council on Education Standards and Testing was created in response to the demands of state governors for the development of national standards and assessments (Campbell & Romer, 1992). The report supplements the President's Education Summit held in 1989. Of particular interest were goal 3, on student achievement and citizenship, and goal 4, on science and mathematics education. For goal 3 the issue was how to renew and enhance civic culture in ways that enable all citizens to participate more effectively in the processes of democracy. Here we need to remember that we are living in a science/technology–oriented democracy. Goal 4 is a statement regarding science and mathematics standards to be met by the year 2000. In goals 3 and 4 we have a rationale for transforming school science curricula.

In the science education reform movements of the twentieth century, children have somehow been forgotten. Overlooked is the fact that one-third of a child's education between birth and high school graduation occurs at home. A general awareness of this situation is noted in the National Education Goals Report (National Education Goals Panel, l991). Goal 1 states that by the year 2000, all children should enter school ready to learn, and goal 8 recognizes that every parent is the child's first teacher and should devote time to helping his or her children learn (National Education Goals Panel, 1995; Department of Education, 1996).

Goal 8 stresses the importance of parental participation in the education of children (National Education Goals Panel, 1995). The goal proposes that "by the year 2000 every school will promote partnerships that will increase the parental involvement and participation in promoting the social, emotional, and academic growth of children." All of the national goals are in one way or another in the context of building a nation of learners.

A report to the U.S. House of Representatives on education in 1991 recognized early childhood as "the cornerstone to quality education." It was recognized that as yet, "we do not have plans ensuring a smooth transition from early childhood education to the elementary school." The U. S. Congress had previously sought to analyze the status and condition of children and family life in America (U.S. Congress, 1987).

The Educational Testing Service (ETS) explored the involvement of parents in fostering children's learning. Barton and Cooley (1992) recognized that family structures have changed radically over the past century, with changes accelerating since the 1950s. Unless we find better ways of dealing with family lifestyles, national educational goals for the year 2000 will be difficult to attain. Schooling in the home has become as serious a problem as education in the schools (Center for the Future of Children, 1995a). Other countries are experiencing the same problem.

In 1995 the Biological Sciences Curriculum Study published a report sponsored by the National Science Foundation titled *Designing the Science Curriculum: A Report on the Implications of Standards and Benchmarks for Science Education* (BSCS, 1995). The essence of the report was that after a decade of dreams, efforts to achieve major changes in science education by the year 2000 do not seem likely.

The U.S. Congress has played an active role in fostering educational changes. For example, Senate Bill 3095 (1990) authorized the creation of a National Education Report to measure the educational achievement of both students and schools. In 1991 Senate Bill 2 was passed, establishing a National Council on Educational Goals and an Academic Report Card to measure progress on the goals. The final approval of Senate Bill 2 with specific details was published in February 1992.

Senate bill 2114 (1990) was passed for the purpose of promoting excellence in American mathematics, science, and engineering education and to "enhance the scientific and technical literacy of the American public."

In 1992, a bill (H.R. 4323) was introduced in the House of Representatives "to improve education for all students by restructuring the education system in the States." In 1993 the Senate and the House passed an act (H.R., 1804, 1994) "to improve learning and teaching by providing a national framework for education reform." The official title of this bill is *"Goals 2000, Educate America Act."* The goals cited in the act are those developed at the education summit called by President Bush and attended by all the state governors. The goal of the United States to be "first in the world by the year 2000" in science education was again emphasized.

THE MAJOR ISSUES IN THE REFORM
OF EDUCATION IN THE SCIENCES

To identify the major issues of science education as they are perceived in national reports, a normative analysis was made of the references cited in this chapter. The only additional publication used was *Community Update*, an ongoing monthly newsletter of the National Education Goals Panel. The resulting analysis provides a conceptual or philosophical framework for transforming science education. This framework also provides criteria by which the validity of goals and instructional practices can be assessed. The purposes and effectiveness of proposed goals and practices will undoubtedly require new forms of educational research (Hurd, 1971b, 1983. 1993). The findings of this normative analysis follow.

The research procedure for this analysis is not easily described. I noted the advice of Johann Wolfgang von Goethe, who wrote in the late 1700s:

> It is extremely difficult to report on the opinions of others, especially when they closely agree, border and cross one another. If the reporter goes into detail, he creates impatience and boredom; if he wants to summarize, he risks giving his own point of view; if he avoids judgments, the reader does not know where to begin; and if he organizes his materials according to principles, the presentation becomes one-sided and arouses opposition, and the history itself creates new history. (quoted in "Vignettes: Retelling," 1993, p. 390)

Researchers in social and economic fields caution that the major problem of normative analysis is one of bias. My major bias is the hope that a repeat of previous science education reforms can be avoided. My other bias is to stay clear of science education reform proposals contradictory to a democratic society and the welfare of young people and the current practice and ethos of science.

With these cautions in mind, a summary was developed of national recommendations for the reformation of an education in the sciences with the hope it will be different from today's science curricula. These proposals follow in an abridged form.

1. Education in the sciences should harmonize with life as lived in the real world. Currently students experience their world only after the final bell for the school day has rung.

2. Curricula need to be invented that represent the strategic nature or mission-oriented research of contemporary science. The traditional discipline-bound, science career–oriented courses are too narrow in scope to serve as a base for a citizen's education in the sciences.

3. The educational process for today's knowledge-intensive society needs to begin soon after birth. Goal 1 of the National Education Goals Panel states that children should enter school ready to learn. The family and community bear this responsibility, assisted by schools. We need to recognize that in the span of years from birth to high school graduation, one-third of the learning time is in the hands of the family and the community (Department of Education, 1996). China and Japan have established planned curricula, under the supervision of parents or grandparents, starting at ages 28 months and 30 months, respectively.

4. An education in the sciences should be in terms of the fulfillment of life, interconnecting the sciences, technology, society, economy, individual development, quality of life, and civic responsibilities. Most current science curricula consist of a chain of facts from page one to the last page of a textbook and fail to meet the educational demands of our changing culture in either purpose or subject matter.

5. Science in the context of life and living recognizes the biological and social developmental levels of individuals from birth throughout life; the reform is focused upon making science more productive in the life of students.

6. Congress, in establishing science and technology as an integral part of our democratic society, makes enculturation a new purpose for the teaching of science.

7. Meeting the science education demands for the twenty-first century requires an extensive study of the current nature and practice of science. This science has no fully accepted name as yet, but descriptives such as technoscience, sci-tech, and applied technology are used to indicate something of the nature of today's science.

8. School science curricula should be organized in terms of problems that connect science/technology with self, community, society, and the future. This is a curriculum beyond the limitations of traditional disciplines and represents the new civic dimension of science education in helping shape the nation's social and economic policies (Morell, 1996).

9. A "science-for-all" context includes a focus on the preparation of all citizens for jobs in our knowledge-intensive world. Today the economic worth of individuals depends upon their ability to acquire, process, and utilize information in different ways. These abilities are different from the traditional concepts of vocational education. Increasingly today, non-knowledgeable persons are being replaced by robots.

The new faces of work are described as "learning a living," "thinking for a living," and "building human capital." Closing the gap between school and work has become a major goal of science education (Jennings, 1995). The School-to-Work Opportunities Act (H. R. 2884) was passed by the 103d U. S. Congress on May 4, 1994. Its purpose was to establish a national framework for the development of opportunity systems for preparation for work and other goals (School-to-Work Opportunities Act, U. S. Congress, H. R. No. 2884).

10. Social inquiry supplements scientific inquiry in importance as a goal for science teaching. Scientific inquiry is discipline bound and has little use beyond the classroom. Social inquiry is a process of utilizing science concepts for resolving personal, social, and economic actions. Beyond the laboratory, science concepts take on a different meaning (Holton, 1975).

11. Laboratory work in the framework of the national science goals is seen as an experience in citizenship. The problems selected for study typically require teamwork characteristic of most scientific research today. Team study of a problem requires developing communication skills essential not only for work but also for fully participating in a democracy. A modern perception of the science laboratory is that it has no intellectual walls.

12. In science education a perspective of the future is seen as essential not for predicting the future but for shaping it (George, 1996). This approach is in accord with the way strategic research in the sciences is oriented. The effort is to develop a science curriculum characteristic of the world in which the student is likely to live (David and Lucile Packard Foundation, 1995; National Science Foundation, 1996; National Center for Education Statistics, 1996).

13. To achieve an education in the sciences to meet national goals will require in a large measure a national curriculum framework. A central purpose of this curriculum is a citizen's understanding of a science and technology–oriented culture and democracy.

14. A national science curriculum framework is viewed as an integrated core subject representing the interdisciplinary nature and blending of con-

temporary research in the sciences. It would vary in emphasis with the developmental level of the student, changes in the practice of science/technology, and current socioeconomic conditions. Science as a core subject is also viewed as a way of connecting the natural sciences to the humanities and social sciences.

15. The nature of knowledge and its relationship to ways of knowing and understanding is still being debated. There is agreement that the goal should be the ability to utilize science knowledge appropriately in resolving problems associated with human welfare and the common good.

16. The assessment of learning would focus on the student's ability to manage science knowledge in terms of problems and issues one is likely to encounter throughout life. The extent to which science knowledge is usable in everyday affairs is a measure of human capital. By the year 2020 it is expected that almost all the knowledge ever discovered will be available to anyone who knows how to identify, access, process, and utilize the information.

17. National reports on science education stress that it is the quality of science curricula that counts. Quality is defined as a contemporary view of science/technology in terms of its meaning for the welfare of individuals and the social and economic progress of the nation. The National Research Council (1979) asserts that the critical goal of science education is "knowledge useful for one's own well-being and knowledge useful for good citizenship."

18. The National Science Foundation notes that current "school science education seems to lack a sense of direction, theory, and philosophy that would provide guidance to curriculum development and instruction. What students should learn also remains unclear" (1980). See also the NSF report to the National Science Foundation Board (1983a).

19. Science/technology in personal and civic contexts requires special ways of thinking, recognized as higher order thinking skills. To achieve this goal requires that students be able to distinguish evidence from propaganda, probability from certainty, relevant questions from pseudoquestions, rational beliefs from superstitions, data from assertions, science from myth and folklore, credibility from incredibility, sense from nonsense, fact from fiction, and theory from dogma. Higher order thinking skills are related to the optimal use of science knowledge in personal and social contexts. Higher order thinking skills are *qualitative* in nature, in contrast to the notion of scientific inquiry, which is *quantitative* and discipline bound.

In addition, when science/technology information is brought into contexts where it serves people and society, elements of ethics, values, morals, bias, prejudice, politics, risks, judiciary, ideals, trade-offs, and probability become a part of the thinking process. There are also times when a problem may have more than one "right answer." Problems of this sort call for laboratory experience in classical debate.

Problem solving, decision making (or no decision), logical reasoning, and forming judgments and actions are the end products of higher order thinking skills. These skills are also viewed as a way of converting science information into knowledge, as well as providing a base for the "learning to learn" purpose of science instruction, which in turn is a way of putting students in command of their own intellectual potential throughout life. Higher order thinking skills are required to make science information more productive in human affairs and for advancing the nation's economy (Hurd, 1989b).

20. The proposed view of school science matches the nature of contemporary science, with its emphasis on strategic research, designed from the onset to benefit human well-being or social or economic progress. About 75% of the research in the sciences is now identified as strategic or mission-oriented research. In terms of science education, the trend is also described as relating science to the real life or real world of the student. In this context the student *is* the curriculum. What is sought is a curriculum that can be experienced and lived by the learner for life in a changing world.

21. Over the past several decades and continuing is the development of the cognitive sciences. Cognitive scientists investigate how human beings learn, remember, and utilize knowledge. What interests the cognitive scientist is a view of how to foster an understanding of science and the optimal utilization of this knowledge in the contexts of science and society. It has long been recognized that a major outcome of conventional science courses has been that of forgetting. Now that knowledge has become the basis of one's economic success in life, a measure of one's social capacity, and the principal treasure of our civilization, learning in the sciences takes on new meaning. Some biologists view the birth of a knowledge-intensive society and its influence on human adaptive capacities as marking a new phase in human evolution, a move toward *Homo sapiens sapiens*, the product of cognitive adaptive capacities.

22. In an ever-changing knowledge-intensive world, the human mind must constantly be refueled with new information of the proper sort.

A National Research Council (NRC) report in 1978 points out the difficulties of relating knowledge developed in the natural sciences with that produced in the social sciences. The report deplores the "sluggishness" of discussion on this issue (NRC, 1978). A modern education in the sciences is seen as one that helps connect students with the natural world, the culture, work, society, and most of all, oneself. All of these factors are interconnected in various ways (Hurd, 1970b).

23. New assessment and testing practices will be required to harmonize with the new goals and modes of thinking proposed for modern sci-

ence curricula. Traditionally, tests have been used to determine a student's reservoir of information on a topic. All students take the same test. Grading is a matter of determining winners and losers at a cutoff at some percentage of right answers.

Assessments being sought for the modern curricula are those which recognize every student as a variable. The purposes of the new tests are to indicate the capacity of a student to utilize what has been learned in ways appropriate for responsible living in a knowledge-intensive society.

The results of a test should provide students with insights on how to improve their ability to utilize what they have learned. One's competitive position is with oneself. These tests seem possible with the aid of computers. A debate on national testing is found in the report of the National Council on Education Standards and Testing (NCEST, 1992; Hanushek & Jorgenson, 1996).

24. The starting point for a reform of science education should be a study of students and the problems they are facing in this transition period to a new culture. Youths and family structures today are different from those of a generation ago. More live in poverty, more are homeless, more commit suicide, more lack the benefit of health care, and more are having difficulty adapting to a changing society and understanding the changing world of work and themselves (U. S. Congress, 1983, 1987).

The new biological sciences have much to offer for alleviating the problems of today's youth. For all of the twentieth century every science curriculum reform movement has included a goal of "meeting the needs of students"; none of the previous reform movements ever delivered the appropriate curriculum. A reform task today becomes one of inventing new science curricula that relate youth to themselves, their moral and political obligations, and their social responsibility for living and participating in a knowledge-intensive society and for understanding a science/technology–oriented democracy.

25. Throughout all topics in science courses there should be the concept of change. The sciences are dynamic fields of study with an "endless frontier." Students are misinformed when they do not recognize that the topic they are studying today is likely to be different tomorrow. Contrast your knowledge of the universe before the Hubble Telescope with your knowledge of astronomy today. Twice since I started to write this chapter the universe has "increased" in size. A primary purpose of science education has become one of connecting students to a changing world.

26. The tone of national efforts for the reform of science education is an integration of science with other school subjects in ways that will increase opportunities for critical thinking and social interaction. The ultimate goal is to expand the interdisciplinary characteristics of contemporary science/tech-

nology with social and economic development in ways that recognize that the wealth of a nation and of an individual today are determined by usable knowledge. Current science curricula are mostly a dead end in this context.

27. The National Education Goals Panel sees the need to coordinate all community agencies and others who have a concern for science education. These agencies include museums, zoos, parks, aquariums, environmental groups, businesses, hospitals, parents, and literally hundreds of others in addition to schools (ASTC Newsletter, 1995; Smithsonian, 1996). Efforts to bring about perceived changes essential to a new world order require education reformers and teachers to think in new ways from reformers earlier in the twentieth century.

RESHAPING SCIENCE EDUCATION
FOR A CHANGING CULTURE

Currently there is no way to summarize all the changes expected in the reformation of an education in the sciences. We are entering a new era in our national life, as well as experiencing revolutionary changes in the nature and practice of the sciences. Furthermore, all these forces are interactive. There is agreement that all students should be educated for full citizenship in our science/technology–oriented democracy and for a higher quality of life. To achieve this, students must acquire learning skills and incentives that will motivate them to keep their minds actively engaged in learning throughout life. There is no other way to cope and adapt in a knowledge-intensive society. It is the interconnectedness of these attributes that provides a base for a new vision of science education and (one hopes) the development of a theory or philosophy of science education from which the viability of reform efforts can be judged. The array of educational policies that now exist fail to recognize the new dimensions of science and technology in human and social affairs (Hurd, 1993).

THE PROBLEM APPROACH TO SCIENCE EDUCATION REFORM

Throughout this century most efforts to improve education in the sciences have been to determine the problems that beset schools and then seek to resolve them one at a time. In terms of a new curriculum, the first step is to update the subject matter of the existing curriculum in terms of relevant science disciplines. Academic scientists are asked to take on this responsibility (AAAS, 1993; NRC, 1996). At other times, existing courses are rearranged in the school curriculum for the purpose of improving learning by recognizing levels of intellectual developments (NSTA, 1992).

The public seeking to bring about a reform of schooling does so by passing laws and regulations regarding the conduct of education. It is estimated that since the 1970s, close to a thousand state laws and local mandates have been passed to force schools to change. Typical actions include increasing the number of science courses required for graduation from high school, demanding that teachers have more science courses for certification, lengthening the school day and year, redefining the responsibilities of school principals, more testing of students as well as of teachers, requiring more laboratory activities, making schools more effective by hiring noncertified business-trained people to run schools, judging student and school achievement by standardized tests, barring students from athletic teams unless their grades are at or above the school average, demanding that schools get back to basics, and making science courses more rigorous (generally interpreted as rugged). In general, parents support the need for reform but not for change.

Every curriculum reform movement of this century has been paced by a new technology which, if used properly, was assumed to improve learning. Schools began this century by making wider use of blackboards, then came the use of video slides of various sizes, then radio was introduced to provide an ongoing ear to the world of science, followed by 16mm and then 8mm films. The invention of television was viewed as providing an eye to world events in science. CD-ROMS and now computers provide new ways of managing knowledge. The ongoing invention of new laboratory equipment seems endless. There is little question but that all these developments can in some way improve teaching, but to what end is questionable.

From time to time, additions are made to existing science curricula, such as one- or two-week units on human reproduction, AIDS, youth violence, local environmental studies, or health topics. Teachers typically view these projects as add-ons that often interfere with completing the textbook by June.

Many schools seek curriculum improvements by forming alliances with businesses, colleges or universities, professional science societies, or parents. It is not a question of whether these alliances are worthy, but that a clear purpose is usually lacking; for example, how do they enable students to find their way in a knowledge-intensive society, and do these projects advance a science/technology–oriented democracy?

Typical efforts to reform education in the sciences have been guided mostly by tradition. Curriculum development has been a matter of tinkering in efforts to reorder, restructure, reshape, revise, and refine past efforts. By contrast, the U.S. government sees the problem as one of inventing new curricula to deal with problems associated with our changing culture, the new image of science and technology, more productive ways of knowing for a knowledge-intensive society, and the changing characteristics of students (Hurd, 1970b). What is required are entirely new policies recognizing the changing nature of science, society, and students, and the new responsibilities of schools in a science/technology–oriented democracy.

Both the national policy and the problem groups of reformers suffer from a failure to recognize the historical perspectives on a topic. For example, the emphasis on "hands-on science" was first questioned in the 1850s by Louis Pasteur, who pointed out that children will perceive from an activity only what they have been prepared to perceive. Another example is the emphasis on the use of portfolios for evaluating student achievement. In the 1930s they were profiles and had a short life. They proved to be too time consuming for teachers; also, statisticians opposed profiles because they could not be interpreted statistically.

The isolated and self-contained nature of science reform projects that now exist is not likely to lead to the reformation of science education that is being sought. Yet it is these projects that have attracted the attention of most professional science educators and school administrators, especially the actions that are visible to the public.

THE TASK AHEAD: DEVELOPING
A THEORY OF SCIENCE EDUCATION

How to integrate the shifts in science, technology, society, and culture requires a change in the way we think about education in the sciences. The public interprets these demands for educational change as a crisis. A crisis represents a perceived disturbance in a social system that ought to be dealt with. Our "crisis" is a science for everyone in the context of a knowledge-intensive age. Researchers in science education need to concentrate more research on society in relationship to the economy and the quality of life. These are not issues that can be dealt with statistically.

One needed aspect of research in science education is a consideration of the future. While the future may not be fully predictable, reform issues can be dealt with in terms of a "rational future." Robert E. Lucas, Jr., won the 1995 Nobel prize in economics for his development of the notion of "rational expectations" for the economy in the years ahead. The same concept applies to science education in attempts to release it from tradition and its mismatch with today's world, especially the contemporary nature and practice of science. The public is becoming impatient with so little progress. The U.S. Congress views the task of educational reform as follows:

To improve learning and teaching by providing a national framework for education reform; to promote the research, consensus building, and systematic changes needed to ensure equitable opportunities and high levels of achievement for all American students; to provide a framework for reauthorization of all federal education programs; to promote the development and adoption of a voluntary national system of skill standards and certifications; and for other purposes. (H. R. 1804, 1994, p. 1)

Problems and issues in science education have dimensions in society, culture, cognition, human development, and the common good. Until recently the qualitative nature of this research has served as a barrier to its publication in science education research journals. The *American Education Research Journal* (Urban, 1990) appointed an editor and has established a section for the publication of research on "social and institutional analysis."

The importance of developing a theory of science education is to provide a framework in which elements of reform may find meaning and viability. It provides a means for making consistent choices likely to improve the course of reform. In other ways it makes it possible to distinguish buzzwords, slogans, postulates, and clichés from rational thought. It also provides a base for debates, deliberations, and reflection on problems in place of "instant philosophies" and "off the top of the head" responses.

First steps to bring the national curriculum standards into a goal-oriented framework have been made by a National Research Council (1996) committee chaired by W. R. Bybee. The report considers the curriculum within the perspective of scientific literacy for all students (Bybee, 1996). A second study by a National Science Teachers Committee report edited by Juliana Textley and Ann Wild frames the national curriculum standards in terms of a scientific inquiry goal (Textley & Wild, 1996).

Both studies represent an examination of past efforts to reform school science education. The NRC standards are essentially an updating of the progressive education movement of the 1930s and 1940s, when considerable research was done on the "principles and generalizations" of science in a science-social context (Thayer, 1938, pp. 142–344). The NSTA emphasis on scientific literacy as a curriculum framework amplifies reports by Hurd (1958, 1970b).

CHAPTER 3

Roots and Evolution
of Science Education

During the entire history of the United States, efforts have been made to relate teaching of the natural sciences to individual welfare and social and economic progress. In fact, this view was held by Francis Bacon, who first used the term *modern science*, when in 1620 he wrote: "The ideal of human service is the ultimate goal of scientific effort." Bacon proposed that the subject matter for science education be "that which has the most value for the welfare of man" (quoted in Dick, 1955, pp. 141, 147).

In 1743, Benjamin Franklin founded the American Philosophical Society with the primary purpose to "promote useful knowledge and improve the scientific experience, observations, and experiments, which if well examined, pursued, and improved might lead to discoveries to the advantage of some or all of the nation's plantations and to the benefit of mankind in general" (Franklin, 1743).

While he was Vice President of the United States in 1798, Thomas Jefferson noticed little "practical" science was being taught in the schools. Jefferson viewed "sciences as keys to the treasures of nature . . . hands must be trained to use them wisely" (quoted in de Nemours, 1923). Jefferson saw the major purpose of education in the sciences as one of enhancing the progress of a developing nation. A survey of U. S. schools by Dupont de Nemours revealed that Jefferson's views were not reflected in the science textbooks schools were using. Jefferson asked Congress for money to rewrite science texts to reflect "natural history and mechanics," but Congress refused the money on the grounds that school curricula were options of the local community. The current proposals for national science standards and goals in science have again raised questions about the place of the federal government in setting school policies.

Jefferson's views were not entirely lost; they found a place in several new universities. Rensselaer Polytechnic Institute was founded in 1824 with the purpose of relating "science to the common purposes of life," as well as to foster "the application of experimental chemistry, philosophy, and natural history to agriculture, domestic economy, the arts, and manufacturing" (Eddy, 1956, pp. 13, 37).

In 1847, James Wilkinson, a member of the Royal College of Surgeons of London, delivered the now famous lecture "Science for All." Wilkinson expressed the view that "the end for which knowledge was sought and recorded by the learned, and the end for which it is required by the multitude, are not the same, but different ends," especially in science (1847, p. 3). He noted that scientists consider scientific knowledge as intellectual property to be transmitted "unimpaired from generation to generation, rather than farming it out with a simple regard for public service" (p. 4). Wilkinson stresses the point that scientists want to be judged by their peers and are unconcerned with relating their findings "in the business of life."

The 1850s were a period of reexamining the teaching of science in schools, not only in the United States, but also in England. An insightful essay, written in 1859 by British philosopher Herbert Spencer, was titled "What Knowledge Is of Most Worth?" Spencer thought whatever was chosen to be taught should "contribute to personal well-being and have a bearing on some aspect of life." He viewed school science courses as consisting mostly of "dead facts" that fail to make clear they they can produce appreciable effects on human welfare. He saw the subject matter of science curricula as focused on knowledge "which brings applause to scientists ignoring knowledge conducive to personal well-being and how to live in the widest sense." He interpreted the "widest sense" as including self-preservation, citizenship, and the refinements of life as one prepares for complete living (1859, pp. 14–15).

In 1861, a committee of Boston citizens founded a new university, the Massachusetts Institute of Technology. The committee felt the time had arrived to consider "the happy influence of scientific culture on the industry and civilization of nations . . . and for the cooperation of intelligent culture with industrial pursuits," recognizing that "material prosperity and intellectual advancement are inseparably associated." The committee also developed a rationale, curriculum goals, and instructional strategies for what would today be identified as a science/technology society curriculum (Committee of Associated Institutions of Science and Arts, 1861).

In 1862 the U.S. Congress passed an act "to donate public lands to the several states and territories which may provide colleges for the benefit of agriculture and mechanical arts." The legislation became known as the Morrill Act. Each state accepting free land was obligated to establish

> at least one college where the leading object shall be, without excluding other scientific and classical studies, and including military tactics, to teach such branches as are related to agricultural and the mechanical arts . . . in order to

promote the liberal and practical education of the industrial classes in the several pursuits and professions of life. (Eddy, 1956, p. 10)

The act made it clear that higher education must be changed to recognize advances in science and their crossover with agriculture and the mechanical arts.

The purpose of the Morrill Act was to provide a collegiate education in the sciences that would be useful for those "who use it and closer to the needs of the average man." The fundamental idea was to

offer an opportunity in every state for a liberal and larger education to large numbers, not merely to those destined to sedentary professions, but to those most needing higher instruction for the world's business, for industrial pursuits, and for the profession of life. (p. xiii)

The intent of this act was to provide students with a practical education, one that they could experience. It took close to half a century to develop the envisioned collegiate programs, delayed mostly by the lack of faculty members who knew practical uses for what they taught and who could go beyond theory. Within the next century, programs in vocational agriculture, including mechanics, were to be found in a majority of high schools.

Major changes in both science and our society characterized America from 1890 to the 1900s. In that period of time the nation began a rapid shift from an agricultural society to an industrial one. There also emerged a new era of science typified by major advancements, especially in the physical sciences. The new industrial society created a demand for skilled workers who were better educated than traditional workers on farms and in factories.

It soon became apparent that the education being offered by the schools in the 1890s was out of step with both science and the emerging industrial society. The National Education Association (NEA) in 1892 appointed a Committee of Ten to study changes needed in schooling. A subcommittee composed of teachers, educators, and scientists was appointed to make recommendations for improving education in the sciences. Later the science committee was made a permanent section of the NEA, the first nationwide science teachers' association.

After extensive deliberations over a two-year period, the science committee (NEA, 1894) recommended that all students be required to take two years of science to graduate from high school, one each in the biological and physical sciences. The rationale underlying this requirement was that these courses would prepare students for the "duties of life." The purposes of science education were seen as imparting information, training students in the powers of accurate observation, memory, and communication, and developing their skills in reasoning and logical investigation. The purpose of laboratory work was seen as "finding a relationship between facts and laws." The committee emphasized that "there should be no difference in the treatment of physics, chemistry, and astronomy for those going to college or scientific schools and those going to neither" (NEA, 1894).

The NEA recommendations were widely discussed during the early years of the twentieth century and debated in a variety of science committees representing biology, chemistry, physics, and physical geography. Ten years after the NEA recommendations were made, no significant changes were to be found in school science textbooks. During this period school enrollments increased, but the percent of students electing science courses decreased. By 1915 it became apparent that something needed to be done to improve enrollments in science courses. New courses in science with a more practical emphasis were developed, such as "household chemistry and physics," as well as special courses called "girls' physics" and "girls' chemistry." Ecology and environmental topics became a part of biology courses. In all such courses the technological applications of science were a feature.

The new course "general science" was designed in 1915 to be offered in the eighth or ninth grade. Its purpose was to excite student interest in taking more science in high school. The course sought to relate science and technology by including topics on aircraft, automobiles, steam engines, radio, telegraph, and electrical household appliances. Although general science became widely popular with students, it failed to create much interest for taking additional science courses.

In 1913, the U.S. Bureau of Education appointed a committee to make recommendations for the reorganization of science in secondary schools and to do so in terms of the needs of pupils and of society (Caldwell, 1920). After seven years of deliberations the committee recommended a science program that would erase the dualism between the practice of science and a citizen's education in science. They perceived the traditional courses in science as being remote from human experience and welfare, neglecting connections with society—home, farm, industry, and students personally. New science courses reflecting these goals did not emerge.

In the United States the 1930s were a decade of great social tensions characterized by a severe economic depression and an unemployment rate that exceeded 20%. Science and particularly technology were seen as the root of all problems. Farms were being mechanized by the use of tractors, dairies by milking machines, and cotton crops by picking machines. Southern sharecroppers could not afford this equipment and as a result lost their farms. This led to a mass migration of people to large cities to seek employment, but with their lack of technical skills they had few opportunities. Factory work was also increasingly becoming automated, and this decreased the number of unskilled workers needed. Because of the lack of job opportunities, enrollments in secondary schools rose dramatically.

The sum of these factors brought into question the whole of precollege education and particularly the teaching of science. The U.S. Commission of Education appointed a committee to study the prevailing goals of science teaching. The committee found that science teachers were not particularly aware of any goals for teaching science beyond those related to the nature of a science discipline. Teachers identified themselves specifically as instructors of physics or chemistry or biology—not as science teachers (Beauchamp, 1932).

In 1932, the National Society for the Study of Education appointed a committee to examine the role of education in the sciences. The committee concluded that the purposes of science teaching should include:

1. Contributions of science for life enrichment through participation in a democratic social order
2. A curriculum that contains "the principles and generalizations of science that ramify more widely into human affairs"
3. Courses organized into units that represent major problems of everyday life and provide opportunities to utilize science in one's own life experience
4. Laboratory work that is an integral part of problem solving
5. A process of learning that involves the integration of facts and experiences leading to the development of ideas. (Powers, 1932, pp. 42–43)

The specific impact of technology on human affairs was not considered. The immediate effect of this report was to stimulate efforts of a number of science education researchers to identify the principles and generalizations of science most common in human experience. In the 1990s these attributes are called "standards."

In 1932 another committee was appointed by the Progressive Education Association (PEA) to study "the educational processes and goals relevant to the needs of learners as they interact with their social medium in situations which confront young people in the home, school, community, and the wider social scene" (Thayer, 1938, p. v). After six years of study and deliberations, the PEA defined new goals for science teaching reflecting the ethos of science, technology, and personal and social needs. The new objectives were classified in terms of (a) personal living, (b) immediate personal-social relations, (c) social-civic relationships, and (d) economic relations. The committee recognized that the elements of problem solving and critical thinking in a science-social context would need to be different from those of scientific problem solving (Thayer, 1938, p. 59).

Whatever progress was made in rethinking science education came to a halt with the advent of World War II. The scientific community pressured for a return to the "fundamentals." By a fundamental approach, scientists meant the teaching of science representing the structure of individual disciplines and the basic laws and theories that determine their character.

A Harvard University report in 1945 rekindled the debate on the place of science in secondary schools (Buck, 1945). The committee stated its position as follows:

> Science instruction in general education should be characterized by broad integrative elements—the comparison of science with other modes of thought, the comparison and contrast of the individual sciences with one another, the relations of science with its own past and with general human history, and of science with problems of human society. These are areas in which science can make a lasting contribution to the general education of all students. Unfortunately, these are the areas slighted most often in science classes. (p. 158)

The committee viewed the student's own way of life and personal relations to the immediate environment to be the most critical integrative elements of the curriculum. They also recognized that science and technology develop in parallel, each fructifying the other (p. 150).

Following President Franklin D. Roosevelt's death, President Harry S Truman authorized the establishment of a committee to consider "manpower for research" under the chairmanship of John R. Steelman. The committee consisted of representatives from science teacher associations and 15 scientists coordinated by the American Association for the Advancement of Science (AAAS). The committee recognized that the accelerating economic progress of the United States depended upon increasing the number of scientists, technical workers, and qualified science teachers (Steelman, 1947). The Steelman reports led to the establishment of the National Science Foundation (NSF) in 1950. The foundation was recognized as "an investment in knowledge" entailing the improvement of science curricula to engage students in "real science" and with opportunities to "think like scientists" (England, 1982).

President Dwight D. Eisenhower in 1959 appointed a committee of mostly university scientists, chaired by Lee A. DuBridge, president of the California Institute of Technology, to prepare a report on "education for the age of science" (DuBridge, 1959). The committee took the position that not only should schools and colleges foster education for the practice of science, but in addition, they should "sharpen the intellectual capacities and curiosities of each new generation so as to produce citizens and leaders who will know how to use the knowledge and tools to advance social and cultural life" (p. 1).

In 1960 a committee of 80 science teachers plus a half dozen scientists, with the sponsorship of the National Society for the Study of Education, published *Rethinking Science Education* (Barnard, 1960). As a basis for discussion they used a preliminary report of the AAAS committee on the "social aspects of science" (AAAS, 1957). The committee agreed the time had come for a reform of education in the sciences, but there was no agreement on what should be the nature of a society-oriented science curriculum.

Beginning in the late 1950s and throughout the 1960s, NSF financed a massive program of curriculum improvement projects in the sciences and mathematics at every grade level from the elementary grades through high school. (For a detailed description of these science courses, see Hurd, 1969a, 1970a; Hurd & Gallagher, 1968.) At every level research, scientists were the advisers on the nature of the curriculum for each course. Their first objective was to bring the subject matter up to date both in theory and in fact. Each course was typically discipline bound, without reference to personal-social implications. Because their task was to "improve science courses," references to technology were mostly eliminated. except for an isolated example here and there. A major objective was "scientific inquiry," with a focus on learning the skills and processes by which scientists expand their theories and validate new discoveries—the "so-called" scientific method. New laboratory exercises were developed to be more quantitative in nature and to require more individual thinking than the traditional routine exercises that always had a "right answer."

Courses were designed to be more academic and rigorous in nature. In this way precollege science courses would better serve as genetic and motivational screening devices for students entering college entrance and planning majors in the sciences. College and university teachers prized students who had had these improved curricula, primarily for their knowledge of technical vocabularies suitable for exploring their disciplines. From the elementary school through high school, students were expected to be like scientists and to think like scientists. The justification for this position was that if children understood science as scientists knew it, they would find it inherently interesting. In this way the 250-year-old tradition of designing school science curricula as career courses was maintained and strengthened. It was this approach to science teaching that had so disturbed Benjamin Franklin, Thomas Jefferson, Herbert Spencer, and the science curriculum reformers of the 1930s, all of whom had advocated the teaching of science in ways that would benefit the individual and serve the progress of society, especially our science/technology–oriented democracy. For over 200 years research scientists have opposed, blocked, and ridiculed the teaching of science in terms of public understanding, or as part of the nation's gross national product, or in terms of reference to the quality of life.

The NSF-improved curriculum projects were sporadically adopted by schools. After a few years, student enrollments in the NSF-sponsored courses began to drop. In 1970 the NSF Advisory Committee for Science Education recommended that NSF modify its traditional commitment to career science as the primary purpose of precollege science education. Committee members advocated science curricula better suited to *all* students by placing more "emphasis on the understanding of science and technology by those who are not, and do not expect to be, professional scientists and technologists." The committee stressed one overriding goal for the future: "to educate scientists who will be at home in society and to educate a society that will be at home with science" (McElroy, 1970, pp. iii–iv). The committee saw its mission as the "public understanding of science," a goal requiring a "broader conception of what science is and to demonstrate the relevance of science to society."

A goal of science education that is grossly neglected in most science reform proposals is that of economic progress. The relationship of economies to science education reform is highlighted in national policies. The focus is on building human capital and advancing school-to-work intellectual skills and an understanding of the relationship of science/technology to the productive capacity of the nation in a global economy. This relationship between education and the economy and the productive power of workers was first recognized by Adam Smith in his 1776 treatise *Wealth of Nations* (Hanushek, 1994, p. v). The economic relationship to science education was first described by the Progressive Education Association in 1937 (Thayer, 1938, pp. 235–305).

WHAT DOES THIS HISTORY TELL US?

This sketch of the history of thinking about the purposes of science education provides a base for deliberations about what should be the curriculum goals for science as it is practiced today, our changing culture, and the enculturation of students and the fullest development of their potential. Preliminary thinking along these lines is summarized in Chapter 2 in terms of national goals for science education in the twenty-first century. Reflections on these goals and how they compare with those over the past four centuries show a great similarity in perspective, yet to this day the prevailing view and context are seldom represented in education policies, courses of study, textbooks, classroom practices, and standardized tests. Students still feel compelled to ask repeatedly, "What good is this going to do me?" This nontraditional approach to educational issues requires that we think differently about the purposes of an education in the sciences for *all* students.

Our first problem is related to ways we can overcome the barrier, set by researchers in different disciplines, that science teaching should focus on the structure of disciplines and their theoretical components. The history of science education portrays a broader agenda for an education in the sciences, one that goes beyond the walls of the laboratory. This agenda views the sciences as a means for enhancing the quality of life, assuring social progress, benefiting humanity, shaping our future, and assuring economic progress. Most of the research in today's science is strategy oriented and cross-disciplinary; these are the first steps toward a modern science curriculum.

CHAPTER 4

The Changing Image of Science

This chapter presents an overall description of contemporary science and its nature and practice. Since the 1940s, evolutionary changes have been developing in the sciences. These changes are products of World War II, when research in the sciences was directed toward the technological development of more effective instruments of war and defense. The "air age" is an example; jet engines were designed to replace propellers, resulting in greater speed. The "atomic age" is another example in which basic research in a science leads to practical applications, for example, from radioactivity to determining the wearing quality of automobile tires and the location of blood clots in the human circulatory system.

Historians and philosophers of science at that time increased their attention to the study of science, technology, and society, and more recently, the political, judicial, and ethical factors that influence research in the sciences. Over the past half-century the center for research in the sciences has shifted from colleges and universities to industry, where a majority of today's scientists are based. All of these factors and other perspectives have brought about qualitative changes in the organization and structure of contemporary science. These shifts in the nature of science are changing our civil life, and they create new directions for reforming curricula in the sciences.

THE GREAT TRANSITION

The current reform movement in science education has been under way since the 1970s. Developments resulting from the movement until now were the subject of Chapter 2. What has been sorely neglected in proposals for rethinking the purposes and focus of science education is the recognition that science itself has changed in practice, nature, and ethos. It is these changes that have outmoded from the start most of the science education reform policies and curriculum models being generated. The developers have proposed revisions of traditional concepts of science rather than considering the nature and practice of contemporary science.

It is the totality of changes in the contemporary nature of science that led to the challenge to reinvent science curricula from pre-kindergarten through graduate school in universities. At the graduate school level the curriculum debate is centered on how the next generation of scientists needs to be educated to meet the challenges of the new image of science. New roles for the sciences in our culture are considerations of the place of science in the lives of people, their quality of life, or more precisely, the meaning and form of the natural sciences in human and social affairs. Embedded in this context is a perspective of science education that began with Francis Bacon in 1620 and has persisted to this day, as noted in Chapter 2.

A quarter of a century has now passed since it was recognized that a reformation of education in the sciences was needed. The concerns arose from an awareness that the "course content improvement programs" of the 1960s were conceptually not adequate to meet the educational requirements of a changing society, nor did they represent the revolutionary changes taking place in contemporary science. The majority of courses developed during the 1960s have proved to be discipline bound and culturally isolated. Their major purpose was to prepare students for advanced courses and a career in a science discipline, a 200-year-old goal prescribing that school science should be taught as an occupational subject oriented toward the preparation of scientists.

The current pressures for a reform of science education are directed toward a public understanding of science in ways that lead to responsible citizenship in a science and technology–oriented democracy and at the same time bring the natural sciences into the cultural stream of American life. Historically these goals serve to erase the "two cultures" concept of science and society first discussed by C. P. Snow (1959). Essential research on the philosophical implications and a social analysis of these purposes to make the reform movement functional have yet to be completed.

For professional science educators, the totality of these changes calls for rethinking educational policies, devising new methods of research, and evolving a philosophy of science education that has cultural and social validity as well as scientific authenticity. The target of this study is to provide a context and blueprint for reinventing science curricula in ways that will serve students throughout their lifetime and foster the progress of the nation.

CONTEMPORARY SCIENCE AND SCHOOL SCIENCE CURRICULA

The science education reform movement has failed to recognize fully not only the transitions taking place in the lives of people and in our society, but also the nature and practice of contemporary science (SCANS, 1992a, b; *Goals 2000*, 1994). How these factors are intertwined is sketched in the following sections of this chapter.

Reflections on the changing images of science reveal a wide gap between contemporary science and school curricula (Duschl, 1985, 1986, 1988; Hurd, 1984, 1985, 1986, 1987, 1991a, 1991b; Kelly, 1993; Maarshalk, 1992; Martin, 1990; Spector, 1993). There is also a need to recognize that science is in a dynamic state of evolution, greater than at any time in the past 400 years. At the same time, there is an antiscience movement and a questioning of the significance of science and technology for human well-being (Ciba, 1972; Gross & Levitt, 1994; Holton, 1993; Wolpert, 1993). Achievements in the biological sciences, to be discussed later, are questions about values, ethics, and judicial judgments as new dimensions of science (Engelhardt & Callahan, 1978; Longino, 1990; Nelkin & Trancino, 1994).

To reinvent the school science curriculum calls for something more than the selection of subject matter solely in terms of its historical significance for advancing research in an academic science discipline. The issue is the importance and relevance of science and technology in human affairs linked to social goals and responsible citizenship (Berkner, 1964; Gardner, 1983; Hurd, 1990, 1993; NSF, 1983a, b; Powers, 1932; RANN 2, 1976).

Transitions in the Nature of Science

My research procedure for determining a sketch of contemporary science has been to abstract and synthesize relevant articles and editorials in *Science, The Scientists,* and *Issues in Science and Technology* over the past five years. Additional information on the conceptual threads that characterize today's science as it is practiced was obtained from personal notes taken at discussion sessions organized by the American Association for the Advancement of Science (AAAS) and the National Academy of Sciences (NAS).

The natural sciences today differ radically from the discipline concept of the past four centuries which continues to dominate school curricula. No particular event or date identifies the emergence of contemporary science. Bruno Latour (1987) uses the term *technoscience* to distinguish today's science from past notions. Derek DeSolla Price, a noted historian of science at Yale University, described today's science as "applied technology" (Price, 1983). Alvin M. Weinberg (1972) coined the term *trans-science* to describe science in its social and political contexts, as in dealing with environmental issues, controlling atomic weapons, changing the DNA pattern of humans, and political pharmacology. *Sci-tech* is another term used to identify contemporary science. STS is widely used to symbolize contemporary science by recognizing the integration of science in the contexts of technology and society.

Currently there is a need for an accepted term that recognizes the nature of contemporary science in both its qualitative and quantitative dimensions.

Fractionation of Science Disciplines

As early as the eighteenth century, the production of new knowledge within the disciplines of science was becoming overwhelming. Within the current century, the exponential growth of scientific knowledge has led to a fractionation of traditional disciplines into thousands of research fields. Biology, chemistry, geology, physics, and astronomy as disciplines have little meaning beyond a way of cataloging seemingly related school or college courses. To identify a person as a biologist, chemist, physicist, or geologist in the sense that they understand most of a discipline is to speak of a nonentity.

How many research fields exist today is unknown. We do know that whatever the answer is today, it will likely be greater tomorrow. The Library of Congress reports it receives 80,000 printed journals, of which over 29,000 are new since 1979; world estimates exceed 100,000 journals, and the figure is growing. The biological sciences have the largest number of journals, 20,000, excluding the medical and health fields (Chronicle, 1990, p. A13). Each of these research fields has its own flavor, its own technical language, its own way of practice, and its own process for interpreting findings.

Hybridizing of Research Fields

Over the past quarter-century there has been a trend toward integrating fields of research that originated in separate disciplines. Examples of these hybridizations are astrophysics. chemecology, biophysics, biochemistry, bioinformatics, geophysics, laser chemistry, biogeochemistry, molecular biology, genetic engineering, neuroscience, bioelectrochemistry, biotechnology, and hundreds more. The relatively new concept of "earth systems" represents a cross-linking of a variety of fields, including the geosphere, biosphere, anthrosphere, geochemistry, and elements of economics and political policy analysis. Current studies of the human brain focus on biochemistry and biophysics in addition to the behavioral sciences to form the new field of neuroscience. Additional examples of the cutting edges of contemporary science can be found in a National Academy of Sciences report (Greenwood et al., 1992, Vol. 1).

The noun *science* as a singular word has become an inappropriate term for studies related to the natural world; a more accurate term would be *sciences*, in the same way we use the plural for mathematics. For example, a summary report of the 1991 meeting of the American Chemical Society states that "like a species that has moved into open niches, evolved, and diversified, chemistry can no longer be regarded as a discrete scientific field. Its methods, concepts, and practitioners are penetrating virtually every nook and cranny of science and technology" (Kerr, 1991).

THE MYTH OF A SCIENTIFIC METHOD

To fully explore the nature of science today would require that each field be examined separately, and to expect to find a commonality in terms of a scientific method would be a futile endeavor (Bauer, 1992). The operating framework of science today is a mixture of theory, available instrumentation, experiment, intuition, insight, trial and error, and thinking in unconventional ways. Research in the sciences has become more an art than a linear method of some sort. (More comprehensive views of contemporary research processes are found in Gleick, 1988; Grinnell, 1992; Latour, 1987; and Woolgar, 1988.)

The National Academy of Sciences, the National Academy of Engineering, and the Institute of Medicine in a joint meeting outlined the practice of science in the contexts of society and economics (National Academy of Sciences, 1986). The panels perceived the method of science to be a task of:

> choosing issues that require attention, setting goals, finding or designing suitable courses of action, and evaluating and choosing among alternative actions. The first three of these activities—fixing agendas, setting goals, and designing actions—are usually called *problem solving*; the last, evaluating and choosing, is usually called *decision making*.
>
> The abilities and skills that determine the quality of our decisions and problem solutions are stored not only in more than 200 million human heads, but also in tools and machines we call computers. . . .
>
> There are no more promising or important targets for basic scientific research than understanding how human minds, with or without the help of computers, solve problems and make decisions effectively, and improving our problem solving and decision making capabilities (p. 19).

Research in the Traditional Mode

In past years most of the researchers in the natural sciences were located in universities. The research conducted was very much an individual affair by a faculty member with some help from graduate and postdoctoral students. Distinctive achievements were often named after the researcher, for example, the works of Charles Darwin, Frances Galton, Gregor Mendel, Isaac Newton, and Joseph Priestley, and for most of the Nobel awardees prior to the 1940s. University scientists still largely work as individuals because they need to create a paper trail to assure promotion up the professional ladder. Their research is a matter of personal interest or curiosity about some aspect of the natural world. Findings which advance an existing theory or open the door to a new theory are the most valued. The whole endeavor is to produce valid knowledge for its own sake.

The Movement Toward Team Research

The scientific revolution now in progress operates within a framework different from that of the past. One characteristic is the development of team research. Today, 95% of research articles are multiauthored; at the beginning of this century the number was 5%. The 12 most frequently cited scientific papers, worldwide, published in 1991 had an average of 5.3 authors per article, and no paper had fewer than 4 authors (Institute for Scientific Information, 1993, pp. 159, 180). In 1986 a research report on respiratory tract infections listed 193 authors representing 20 institutions. It took 17 years of cooperative experimental work on the part of 440 physicists to provide evidence in 1994 that a "top" quark actually exists. Today the findings of individual researchers tend to be viewed with less confidence than those of a team.

Team research provides a broader and more insightful outlook on problems than is characteristic of single individuals. The formation of a research team varies with the problem. Typically a team is used when a problem requires a cross-disciplinary perspective or involves a mix of the hybridized fields or a convergence of the natural and social sciences. Henry Shaffer, a psychologist at the University of Exeter, England, describes research in cognitive neuroscience as "a meeting place of ideas from computer science, psychologists, linguists, philosophers, engineers, neuroscientists, and others interested in the design of intelligent machines" (Veggeberg, 1993, pp. 7, 15). Research on neural networks and the human brain is moving fields of psychology closer and closer to biochemistry.

An editorial in the April 12, 1996, issue of *Science* (Gore, 1996) was entitled "The Metaphor of Distributed Intelligence." The editorial was abstracted from a talk by Vice President Al Gore at the AAAS annual meeting in Baltimore on February 22, 1996. Gore identified research in the natural and social sciences as a product of "distributed intelligence." He pointed out that the most effective way of solving science-social problems as well as those of the natural sciences is through a pooling of intellectual resources from a variety of fields. He saw this approach to research as a distinguishing characteristic of today's knowledge-intensive society.

Essential members of a research team are computers. They store, search, decipher, organize, and communicate information. Computers serve as a cognitive tool in modern research. Some go as far as suggesting interpretive models for the data that have been generated. A symposium sponsored by the American Association for the Advancement of Science (AAAS) recommended that computers be considered "the third branch of science, along with theory and experimentation" (American Association for the Advancement of Science, 1992). A National Science Foundation publication (NSF, 1991) describes computers as "science's leading edge." Further descriptions of the impact of instrumentations on the advancement of scientific discoveries are found in Brauman (1993) and S. S. Hall (1992, pp. 344–349).

Combined with computers are intelligent robots. Robots assist research teams by carrying out routine or dangerous tasks, such as exploring active volcanoes on the ocean floor. It is anticipated that the first "person" likely to explore Mars will be a robot. In addition, robots are capable of working 24 hours a day, with no time out. Their supervisor, also a member of the research team, is an electronic engineer capable of devising special programs or inventing new technology to increase a database.

Researchers sharing their experiences at AAAS meetings felt that a team limit of six to eight persons was best. Too many researchers on a problem complicates communication and the process of collective reasoning. The cognitive power of a team can sometimes be increased by switching the responsibilities of researchers—for example, assigning a natural scientist to deal with the social or economic aspects of a problem, if for no more reason than to ask what might appear to be naive questions.

Developments in electronic communication systems now make it possible for a research team in one location to collaborate with other teams working on the same problem or a related one. It was estimated in 1994 that there were teams researching AIDS-related problems in 90 countries and involving more than 10,000 researchers. The massive amount of information generated by these teams has led to the development of the science of bioinformatics, designed to organize and reorder data and to focus it at various levels of comprehension (Hoke, 1993, pp. 1, 7).

Instrumentation and the Advancement of Knowledge

Progress in each field of science is more and more influenced by available instrumentation (Brauman, 1993). Since the 1950s about 40% of all Nobel awards recognized researchers whose discoveries significantly aided other researchers in extending their observations of the natural world, such as the use of fiber optics. In 1993 the Nobel prizes were divided equally between those who made significant discoveries about nature and those who created new research tools to explore nature.

Throughout the history of modern science, technological developments have set the pace for new discoveries. Anton von Leeuwenhoek's develop-

ment of the simple microscope (1675) made it possible to investigate a previously unknown world of microorganisms. When Galileo (1609) used a
telescope to explore the sky, astronomy took on new meaning. (The relationships between science and technology will be explored in detail in
Chapter 5.)

The place of technology in the advancement of knowledge about the
natural world has been minimized or omitted in science writings (Lenoir,
1988). Scientists traditionally have portrayed their endeavors as purely a
theoretical intellectual adventure (Hurd, 1994a).

Emergence of Strategic Research

From the beginning of scientific research in the United States, there has been
a conflict between academic researchers and applied researchers. Academic
researchers see their task as the generation of new knowledge about the
natural world and solely for the sake of advancing a discipline. Academic
researchers also see no reason to venture beyond the laboratory in communicating their discoveries to others than their peers. Once their research findings and their theoretical interpretation are verified, they consider their task
completed.

Research on a preconceived problem such as AIDS research has traditionally been labeled "applied" science and is held in low esteem by most
academic researchers. Such fields as agriculture, medicine, and technology were forced to establish separate university departments, or colleges
to carry on research from a perspective of public service or human welfare. The ideological concept of pure or basic science as knowledge for its
own sake is now in transition (Webster, 1991; National Science Foundation, 1991).

The new pattern of goal-, problem-, or mission-oriented investigation is
identified as strategic research. Essentially, this is a movement to deliber-
ately extend research beyond the laboratory to the real world of the individual and society—in other words, a move from the esoteric to the exoteric. The idea of strategic research is to factor science problems in terms of
human affairs (Restivo, 1994). Strategic research is a focus on the social relevance of science and with a concern for public utilization (NSF, 1979). The
tentative budget for scientific research supported by the U.S. Congress for
1996 allots 75% for strategic research and 25% for basic research in the traditional sense. The rationale underlying the notion of strategic research has
a long history going back to Francis Bacon in 1620, who wrote, "The ideal
of human service is the ultimate goal of scientific effort" (quoted in Dick,
1955). The history of this movement was the subject of Chapter 3.

The current thrust to link science and technology to societal goals is reflected in two strands of writing: one is oriented toward the public understanding of science (Layton, 1973; Lewenstein (ed.), 1992; Ziman, 1990), and the other in terms of the social functions of science (Bernal, 1939; Jasanoff, 1994; Webster, 1991; Ziman, 1968). These writings indicate that the purposes of school science teaching should include civic learning for participation in today's science and technology–oriented society, and the enculturation of students for understanding science and technology as a cultural force (Woolf, 1964).

The Rise of Research in Biology

Another dimension of contemporary science can be seen in the shift from the dominance of research in the physical sciences to research in the biological sciences. This revolutionary transition has placed biology at the pinnacle of today's science research fields. At the same time the nature and modes of research in biology have changed, making it possible to view biological events from different perspectives such as biochemical, biophysical, physiochemical, or molecular. Biological research is becoming increasingly more interpretive than theory bound—for example, in using studies in anthropology, the environment, medicine, and human ecology, where data are treated qualitatively as well as quantitatively.

In terms of the current national goals for education in the sciences, biology is closest to the realities of life and the culture. As a result, both the funding of research and findings are often the subject of political and social pressures. An example of public pressure to influence research is that of a federal advisory committee recommending that no funding be made available for studies related to (a) cloning or twinning of embryos for purposes of transfer to the womb; (b) genetic diagnosis for sex selection; (c) fertilization across species, such as mating human sperm with gorilla eggs; and (d) attempts to transfer human embryos into animals for gestation, plus other issues (Fletcher, Miller, & Caplan, 1994, pp. 75–80). These and related issues have made bioethics a new field of study. There are a growing number of bioethics courses in university schools of law and medicine, as well as in departments of biology and religion.

It is biotechnology that is leading the nation into a new industrial revolution that is becoming worldwide. At the turn of this century it was physics that generated the industrial revolution. Now one role of physics is to provide instrumentation and information technology that makes it possible to extend the frontiers of biology continually. (The new view of biology is the subject of Chapter 5.)

RESEARCHERS FOR THE CHANGING PRACTICE OF SCIENCE

It is becoming evident that scientific researchers of the future will need to be educated in new ways if they are to engage in cross-discipline research (Abelson, 1986). At best, instead of being specialized in just one field, they will need experience in several fields. Communication skills across disciplines are essential, as well as the ability to think as a member of an integrated cognitive system, the team.

Professional scientific societies have developed programs along these lines. The National Academy of Sciences sponsors its annual Frontiers of Science Symposium, specifically designed to bring young scientists together to listen to other young scientists lecture and to ask the speaker any questions that bother them—none considered stupid. The American Association for the Advancement of Science is increasing the number of interdisciplinary meetings to help bridge knowledge gaps between disciplines. The Scientist to Scientist Colloquium for about 80 participants uses the technique of not revealing in advance each program session's contents so attendees will not skip out whenever a session seems unconnected to his or her work. The goal of all these programs is to help scientists become better research generalists (Moffat, 1993) and more holistic in their views. Science researchers are increasingly engaged in research that goes beyond the spectrum of the natural sciences to include social, economic, behavioral, and cognitive sciences (Spiegel-Rösing & Price, 1977). The education of precollege teachers of science should also include these dimensions. There are programs under way in colleges and universities to develop new ways of educating K–12 science teachers and academic science researchers. A common element in both these programs is the relationship of science to social progress, cultural understanding, and human welfare (Ayala, 1989; Pabst, 1994; Rubinstein, 1994).

CONTEMPORARY SCIENCE AND THE FEDERAL GOVERNMENT

In 1994 a new era in U.S. science and technology was officially introduced with the establishment of federal policies for directing and financially supporting research. For the foreseeable future the priority for federal support of scientific research is in terms of its relevance to human welfare. A primary long-term goal is the development of human capital capable of advancing the economic strength of America (Branscomb, 1981; Hurd, 1989b).

The Carnegie Commission on Science, Technology, and Government, comprising 150 scientists, political leaders, and concerned citizens, after five years of deliberations issued a series of reports recommending that science and technology be linked to societal goals "in the context of human and democratic values" (Carnegie Commission, 1992b) and that the federal government carry the responsibility for organizing and directing related policies (1993a). Specific directions for exploring science as cultural knowledge are described in publications sponsored by the American Association for the Advancement of Science (Chalk, 1988; Scribner & Chalk, 1977; Teich, Nelson, & McEnaney, 1994).

The implications of these studies require a reinterpretation of the purposes for the teaching of science in schools and colleges. The studies cited suggest a new "social contract" between research in the sciences, education in the sciences, and the new realities of living in a nation in which science/technology is a primary cultural force. For researchers in science education, the changes in the relationships of science and society establish a new agenda for educational research.

Science and the Judiciary

Another dimension of contemporary science is its involvement with the legal system. The two cultures, science and law, have their own versions of what constitutes truth; they differ on opinions with regard to scientific evidence and legal evidence concerning an issue. An example is the debate over whether DNA "fingerprinting" has legal validity for identifying each of us as an individual (Hoke, 1994a). And although significant progress has been made in the development of vaccines against major infections, politics and court actions limit their use (J. Cohen, 1994).

Scientists are not always free to pursue research in the manner they feel is most productive. Advocates of animal rights view the use of animals in research as immoral. A majority of scientists, but not all, see this restriction as a threat to public health (Hoke, 1994b; Stoller, 1994). The legal aspects of science in society have become another field of study. The American Bar Association issues the *Jurimetrics Journal*, which is dedicated to exploring issues related to science and the law. At the moment the parallel paths of the scientific community and the judiciary do not intersect (Carnegie, 1993b).

As scientific research has moved into social contexts, new issues are being raised, such as "Who owns science knowledge?" The traditional notion that all research findings are public property is now being argued. Since the majority of scientists are employed in industry and work on problems that have potential commercial value as well as significance in advancing scientific knowledge, how is this dilemma to be resolved? Is it ethical to withhold a research report until its commercial possibilities are explored? Research on health medications is frequently found in this situation. Should it be possible to patent a new genetically altered plant or animal? What constitutes "good" science, that which advances theory or offers a new explanation of an event in the natural world, or that which extends human welfare? Attention is now being directed toward the need for a new social contract between science and society that does not leave the nation "at risk."

TRANSFORMATIONS OF SCIENCE AND SCIENCE EDUCATION

There is a move to place both science and science education into a common framework with frontiers in human, societal, and cultural contexts. Such a procedure will demand new ways of learning, thinking, and utilizing knowledge. To achieve these objectives will require a more eclectic view of teaching, more like that of clinical medicine and engineering. Important intellectual skills will be those associated with social inquiry. Social inquiry is related to ways of accessing scientific and technological information and its proper use in personal/social contexts, an aspect of cultural literacy. The national educational goal of "building human resources" for our knowledge-intensive society represents a new challenge to education in the sciences. How to meet this new commitment to learning is being explored in scholarly ways on a number of fronts (Bandura, 1977; Boulding & Senesh, 1983; Druckman & Bjork, 1994; Graubard, 1973; Machlup, 1962; Penner, Batsche, Knoff, & Nelson, 1994; Resnick, 1987; Shadish & Fuller, 1994; Ward & Reed, 1983).

CONCLUSION

The stated purpose of this chapter is to provide a preliminary sketch of science as it is practiced today. The elements described (and there are others) are to set a measure of purpose and validity for a reinvention of school science curricula. To resolve the dimensions of this problem will require cooperative efforts of natural and social scientists, humanists, cognitive researchers, science educators, and teachers working together in a spirit of goodwill and with a recognition of the cultural, social, and science transformations taking place in our nation.

For readers who have a special interest in research related to science education, there is a need for critical studies within contexts that serve to integrate contemporary science with human welfare and social progress. For a lack of such studies, current school curricula in science are being criticized as "obsolete," a "fraud," and "dead end," serving neither students, science, nor society. This point of view is expressed in *A Nation at Risk* by the National Commission on Excellence in Education (Gardner, 1983).

Before new science curricula can be developed, an expanded agenda of educational research will be needed and findings must be integrated with the cognitive sciences in ways that identify compatible styles of teaching and modes of learning (Atkinson & Jackson, 1977; Hurd, 1990; March, 1985; Shymansky & Kyle, 1990). Two dimensions of this research agenda have been sketched in this study—changes in society, and changes in the practice of contemporary science.

CHAPTER 5

Technology and the Advancement of Knowledge in the Sciences

Before the seventeenth century, Egyptian astronomers had verified the existence of 5,000 stars in the heavens and 2 double stars. Galileo's use of the telescope (1609) established forever a place for technology as a tool for advancing human powers of observation and thus deepening our understanding of the natural world. Anton van Leeuwenhoek, with the development of the simple microscope (1675), opened the doors of biology to a previously unknown world of protozoa and bacteria. Over the centuries to follow, technology has increasingly played a major role in setting new frontiers and shaping the image of science.

In the 1990s optical and space telescopes have served to extend the spherical volume of the universe to a billion light years, or about 6 million trillion miles. We now find ourselves living in a universe thirty times larger than it was previously believed to be. Not only are we in a larger universe with over 11,000 galaxies, but our universe is moving at the rate of 425 miles per second in the direction of the constellation Virgo. This cosmic drift raises new questions about our present notions of the evolution of the universe.

In biology the synchrotron enables researchers to begin unraveling the atomic anatomy of biomolecules and peeking at the smallest units of life: atoms. Three hundred years of technological advancements have made it possible today to think both bigger and smaller when exploring the natural world, from the outer limits of the universe to the molecular characteristics of living cells.

For centuries scientists have sought to portray science as purely a theoretical intellectual adventure. The place of technology in establishing new frontiers has been downplayed. Technology as an integral component of scientific research is not found in any of the National Science Foundation science curriculum improvement projects of the 1960s and 1970s.

My intent in this chapter is to present the coaction of instrumentation and theory in the practice of science. While this relationship has always existed, it is only recently that it has become recognized as a crucial dimension in advancing scientific achievement. Alfred North Whitehead, writing about today's science and the production of new knowledge, commented: "The reason we are on a higher investigative level is not because we have a finer imagination, but because we have better instruments" (1972, p. 107). To succeed today, a scientist must keep abreast of available technologies and theory in a given field. The closing section of this chapter will examine the current paradigm shift taking place in science/technology and its implications for the reinvention and modernization of school science curricula.

A CHANGE IN NAME, A CHANGE IN CONCEPT

The term *modern science* was introduced into our language 400 years ago. The purpose was to distinguish what was then a new movement in science that used experimental procedures for verifying knowledge that had been proposed in a theory. The sciences of the time, such as alchemy, astrology, the "magic" of the medicine man, and the religious treatment of diseases, were questioned because they lacked valid supporting data for their assertions.

The recognition of technology as a formulator of science itself was the subject of much discussion following World War II. J. Robert Oppenheimer, in his Arthur DeHon Little Memorial Lecture at MIT in 1947, noted that the time had come for "the repayment that technology makes to basic science in providing means whereby our physical experience can be extended and enriched" (1947, p. 9). In the 1950s he described the relationship of science and technology as "two sides of a single coin."

Technology influences what can be known in the sciences. The craft of science reflected in the embodiment of technology is so consequential as to change not only the image of science but how we view the natural world, as well—so much so, in fact, that contemporary science is viewed by Derek DeSolla Price, a Yale historian of science, as "applied technology" (Price, 1983).

It is apparent that we need a new name, different from modern science, to represent science and technology as an integrated system for the conduct of research. Since 1945, some 40% of the Nobel prizes have been for unique technological achievements such as lasers, fiber optics, and the invention of the polymerase chain reaction (PCR) that will serve to enhance developments in genetic engineering. In his book *Science in Action*, Bruno Latour coined the word *technoscience* to suggest the interactive character of science and technology in contemporary research (1987, p. 9). Philip H. Abelson notes that today "instruments shape research, determine what discoveries are made, and perhaps even select the types of individuals likely to succeed as scientists" (Abelson, 1986, pp. 182–192). The potential of a young researcher for a career in science has become a matter of technological inventiveness as well as academic preparation and intellectual insight.

At the center of revolutionary changes in contemporary science is the computer, with associated electronics, enabling problems to be explored in ways never before seen as possible. With each advance in instrumentation, from abacus to slide rule to adding machine to calculator to computer, the frontiers of research have been expanded and transformed. Complex research problems dealing with several interacting variables that formerly took years to solve were reduced to a matter of months with the calculator and to seconds using the computer. A symposium sponsored by the American Association for the Advancement of Science in 1992 recommended that computers be recognized as "the third branch of science along with theory and experimentation" (AAAS, 1992, pp. 44–64). Technology as the third dimension of science is itself a form of research focused on instruments that extend observations and make it possible to study phenomena either not known or unaccessible to researchers until now, and the ultimate is not yet in sight.

TECHNOLOGY AND THE TRANSFORMATION OF RESEARCH IN THE SCIENCES

How technology influences research in various fields of science is now explored. The examples chosen are intended to provide perspective not only on how contemporary science is practiced, but also in determining how it changes the way we think about the natural world.

At the core of research in modern technoscience are the computer and associated microelectronics. The question has been asked where science would be today if it were not for computers. A major problem in the sciences for over a century has been and continues to be the proliferation of scientific literature. There are now more than 70,000 science journals, over 20,000 in the biological fields alone. Researchers have sought to resolve the information problem by increasingly narrowing their specialization. The current trend toward a more holistic view of research as well as a unification of the natural and social sciences again complicates issues of data management. Computers have come to the rescue where problems are beyond human analytical capacities. Computers can store more information than humans, remember more, recall it faster, and display it in a variety of modes. They can also minimize any prejudice or bias a researcher may associate with the relationship between information and data.

Computers make it possible to network research on a problem from one laboratory to another. Current research dealing with problems associated with the AIDS pandemic includes cooperative efforts and immediate communication by researchers throughout the world.

Computers allow scientists to attack a problem more completely. This is accomplished by connecting a number of computers in parallel to work on a problem in different ways. One such arrangement executes up to 10 billion instructions per second and retrieves information from databases at the rate of 15,000 documents per second. These massive parallel computers are used in the study of high-energy physics and global climate, and for optimizing and modeling information in linear and nonlinear systems. They have also been found essential for advancing knowledge in fluid dynamics and computational chemistry, as well as in neural modeling. For all such problems, large amounts of data are required, especially where events or entities involve enormous assemblies of particulates such as atoms, cells, or organisms. Parallel computers are also useful in designing artificial intelligence programs, as well as in helping robots learn how to assemble objects.

Modern technology has probably had its greatest impact on research in the biological sciences, leading to new frontiers in biotechnology, bioengineering, biophysics, and biochemistry. Not only has technology revolutionized modes of research in biology; it has also generated new techniques for expanding and changing the direction of research.

For hundreds of years the compound microscope has been a major research tool in biology, but it had narrow limits for observing. With the onset of biotechnology in the 1970s it became evident that new research tools would be needed to advance research. The scanning tunneling microscope is one of these new instruments. This microscope makes it possible to observe chemical bonding within cells. A microlaser attached to this microscope enables one to break these bonds one at a time. The headline of the article reporting this achievement was *For the First Time, Chemists Have Seen Chemistry in Action*. An atomic force microscope displays soft-surface membranes down to the molecular level. A microphysiometer can then be used to detect and monitor the response of cells to a variety of chemical substances. New developments in microscopes use beams of sound waves, X rays, polarized electrons, or the nuclei of various atoms for viewing, rather than light waves.

Computers serve to do the mathematical calculations associated with observations by these microscopes and also to visualize the results—for example, in defining the position of individual atoms in a molecule and to predict how certain shapes might dovetail. The latter is a key factor in the design of new drugs. It is expected that with the completion of the human genome project, 10 trillion bits of information will have been generated. A new computer science of bioinformatics exists to study ways of keeping track, deciphering, and ordering this enormous volume of information.

The influence of technology on modes of research is leading to a new orientation in biology based on achievements in biotechnology and genetic engineering. It is anticipated that by the year 2000 biotechnology will be at the center of a worldwide industrial revolution and provide changes in ways we view the natural, social, and behavioral worlds. For the researcher, revolutionary ways are being generated for dealing with the interactions of technology with theory and theory with practice.

The history of astronomy has been one of increasingly more powerful telescopes and supporting technologies. The naked eye was at one time the sole source of information about the heavens. The invention of the telescope extended the range of the human eye, and, combined with photographic plates, made more data available. All new observations require that they be explained and incorporated within existing theory; if they do not fit in, a new theory must be created.

Today new kinds of telescopes on Earth and in space offer a chance to begin answering such age-old questions as: Just how big is the universe? Does it have an edge? What lies beyond the stars? What is the structure of the universe? What are the conditions for life in the universe? How were stars and galaxies formed, and how did they evolve? What are the sources of cosmic energy? New observations of the universe since 1980 are at a pace that may require astronomers and astrophysicists a century to order and interpret. It is hoped that computers will be able to speed up the ordering of these new observations as well as formulate tentative models regarding their meaning.

The following paragraphs describe examples of earthbound and space technologies that are serving to extend our relatively nearsighted views of the universe. A new telescope (Hobby-Eborly), under construction in Texas, is designed to measure the surface activity of stars, determine the distance of galaxies and stars from Earth, discover more about black holes and the last moments of their existence, and make observations for refining theories about star formation and their evolution. Whatever is observed depends upon the technology that will be generated. In any such venture there is no assurance that new technologies will work as expected, but generally, about 50% do. Then there is the question of knowing what is being observed.

The Keck Telescope in Hawaii has made it possible for astronomers to observe and measure tiny wisps of a primordial element located at the far end of the universe as it looked a billion years ago. The telescope has also captured the light of a quasar located 13 billion light-years away. The search is now on for extrasolar planets known as pulsars, and two have been discovered so far.

The waters off the Kona Coast in Hawaii are among the purest and cleanest in the world. A technology is being developed to use the ocean floor as a telescope for studying cosmic neutrinos. Neutrinos presumably carry a wealth of astrophysical information that cannot be reached by ordinary telescopes. This telescope will contain 216 sensitive photodetectors on the ocean floor, 4,800 meters down. If the telescope works as planned, we will know even more about black holes and cosmic phenomena than we know now and astrophysics will take on new dimensions.

At Pune, in northeastern India, a telescope is being built that consists of 30 parabolic antennas each 45 meters across. This Giant Meterwave Radio Telescope will probe, among other things, the primordial gas clouds that condensed into galaxies in the early days of the universe.

Since the 1970s a flotilla of spacecraft have been launched into the sky. Each one has been loaded with technologies to study the universe in new ways and to reveal what has not before been apparent. Following are some findings from these laboratories in space.

Seeing what one did not expect to observe has often been an outcome of space technologies. For example, *Voyager II* found that the planet Neptune has a 16-hour rotation period for its interior. The surprise was that the surface winds blow the "wrong way" related to the interior, toward the east instead of the west, as is the case with Jupiter, Saturn, and Uranus. Also observed was that the magnetic fields on Neptune and Uranus are remarkedly similar to each other's, but they differ from those of other planets. Uranus was found to have 10 web-thin rings surrounding it and 5 new moonlets within the rings. Astronomers have described data gathered on the atmosphere of Uranus as a "patchwork" calling for a new theory for interpretation. So far, *Voyager II* has produced 5 trillion bits of data, including 100,000 pictures of the outer solar system during its first 7 billion miles of exploration. Computers now have the task of ordering this information in ways likely to lead to new theories and explanations.

The *Magellan* spacecraft has been exploring Venus, and particularly its craters. Observations indicate that most of the impact craters pocketing the surface of Venus look untouched by faulting, folding, or volcanism. This observation hints that the surface of Venus has been quiescent for the past 500 million years or so. The finding may also be interpreted as indicating that Venus is dying a slow death, but not all astronomers agree on this interpretation. A new hypothesis is awaited.

The Hubble Space Telescope is the most elaborate and sophisticated of satellites so far developed. During its expected 15-year lifetime, a host of anticipated programmed observations include: (a) clearer pictures of space objects; (b) improved accuracy of previous observations by using computers to remove interfering light; (c) a search for water-bearing asteroids; (d) measures of the extent of our universe and the rate at which it is expanding by looking back in time; (e) determination of the age and size of a mélange of stars, supernovas, galaxies, and other cosmic bodies; (f) more information about black holes, photoplanetary disks around nearby stars, interstellar molecules, the Crab nebula, supernova remnants, and extrasolar planets; (g) calibration of space; and a host of other types of observations.

New technologies aboard the Hubble have resulted in clear images of galaxies more than 7 billion light-years from Earth. For the first time, giant gas rings have been observed around the celebrated exploding star Supernova 1987A. The Hubble Space Telescope has provided a clear image at the heart of the giant Galaxy M87 in the constellation Virgo, where astronomers have long suspected a monstrous black hole exists. Here the gravity is so great that not even light can escape.

The production of specially designed space technologies continues. The Japanese are developing a new type of space rocket to provide more information about weather and lightning. A unique astrophysical observatory in New Mexico is designed to study the most violent events of the universe, such as exploding black holes. Radar from a space shuttle is being used to map the amount and type of vegetation on areas of the earth. The data will be used to assess the amount of biomass and environmental changes.

The science of space is still in an early phase of research, and there is much more to be known and explained about the universe. The discovery of planets existing 7,000 trillion miles from Earth in the constellation Virgo awaits the development of new and more powerful technologies for studying them directly.

Every field of contemporary science is dependent in one way or another on appropriate technology to advance knowledge. In chemistry a technique has been developed by which individual atoms can be used to build custom-made molecules for specific purposes. In turn, computer programs have been developed that make it possible to visualize chemical processes within these molecules.

Scientists with backgrounds in chemistry, material science, biology, mathematics, computers, and engineering cooperate in working on "smart or intelligent materials and structures." These are human-made artifacts that can, like their creators, sense and respond to their environments by learning, adapting, and repairing themselves. Such developments have found expression in the creation of robots and automation processes in general. A laboratory assistant of choice today is an "intelligent robot" much like an automated bank teller. Robots can be used to remove from laboratories radioactive wastes that are too toxic for people to handle. Each is capable of carrying out experiments 24 hours per day with accuracy and reliability, and with a computer as its assistant, it will process on demand the information gathered. Imagine a laboratory assistant who never complains about working conditions, who follows instructions, and who is highly productive.

Technological progress has extended the frontiers of physics over the past half century and has also influenced research in other fields such as biophysics, biotechnology, astrophysics, geophysics, and physical chemistry. Advancements in all of these fields, including the physical sciences and the life sciences, have roots in the findings of particle physics.

Efforts to discover the building blocks of matter have led to the development of giant instruments and a host of new technologies. The largest of these instruments is the electron-positron collider near Geneva, Switzerland. This instrument is 16.8 miles in circumference, 5.3 miles in diameter, and an average of 360 feet underground. It harbors nearly 5,000 electromagnets, 4 particle detectors weighing more than 3,000 tons each, 160 computers, and 4,000 miles of electrical cable. A single experiment requires the combined efforts of hundreds of researchers. Among the many questions it is hoped the collider will help to resolve is what happened 15 billion years ago in the seconds following the big bang that gave birth to the universe and all it contains.

In 1944, physicists at the Fermi National Accelerator Laboratory presented the first experimental evidence that there was a "top quark," the heaviest of all quarks, 200 times heavier than a proton. It took 17 years of experimental work on the part of 440 physicists to provide evidence that the top quark actually exists.

At the Princeton Plasma Physics Laboratory, temperatures of 100 million degrees—three times hotter than the core of the sun—have been used to ignite a mixture of tritium and deuterium gas to bring about a nuclear fusion reaction. This reaction generates nearly 5 million watts of energy, the eventual goal is 10 million. When scientists began working on fusion a half-century ago, they had no idea the process would be so difficult. The process is important because it produces more energy than it consumes. Whether it will ever be commercially feasible is yet to be determined. If so, we could enter a new age in human history.

The discovery of fiber optics and systems that use light as carriers of information offers the possibility of extremely high speed processing, transmission, and storage of data. A host of other new technologies based on the use of photons is generating a whole new era in the study of optics.

The bonds between science and technology for advancing knowledge that have been cited here are intended to illustrate the dynamic character of contemporary research in the sciences. Since the middle of the twentieth century, technology has changed in fundamental ways the nature of scientific research, and it is continuing at an accelerated rate. Technology has served to open new doors to problems that can now be explored, shaped the research that follows, and provided help for interpreting observations. As we have seen, technology frequently delivers unexpected observations that will require new theories or the modification of older theories to provide a valid interpretation. Increasingly, scientific inquiry requires that researchers think in new ways, especially when considering interdisciplinary problems. Traditional processes of inquiry in the sciences have been discipline bound and therefore limited in scope. The contexts of science today, particularly problems that are nonlinear or which have a cross-disciplinary character, do not respond to the traditional modes of scientific inquiry. A new science, *chaos theory*, is being developed to deal with problems that are now seen as disordered, such as weather, tides, turbulence, living populations of animals, and the behavior of nerve impulses. These are among the messy problems into which chaos theory portends more insight than we have now. Currently, chaos theory itself is being questioned.

IMPLICATIONS OF TECHNOSCIENCE
FOR PRECOLLEGE EDUCATION IN THE SCIENCES

School science courses have traditionally been taught as a history of theories and sustaining concepts, constructs, or generalizations. Timothy Lenoir, a historian of science, notes that "experiments, instrumentation, and procedures of measurement, the body of practices and technologies forming the technical culture of science, have received at most a cameo appearance in most histories" (Lenoir, 1988, pp. 5–22). And, one may add, in nearly all school and college science textbooks now in use.

The national effort to modernize education in the sciences has as one of its goals a valid interpretation of contemporary science and its practice. If American students are to meet the national goal of "best in the world by the year 2000," one dimension should be that they have the best understanding of how scientific knowledge is currently being produced. To achieve this purpose will require that a science concept be taught in a coactive context that blurs distinctions between science and technology. It should also be noted that there are only rare times in which technology does not play a part in generating a new theory, for example, the insights of Watson and Crick in developing the double helix model of heredity, Darwin's conceptual scheme of organic evolution, and Einstein's theory of relativity.

Key Features for Understanding Science/Technology
as Currently Practiced

1. Technology expressed in instrumentation determines what ultimately can be observed in the natural world.

2. The progress of science throughout history and increasingly so today is characterized by developments in technology.
3. A skilled researcher today is one who possesses the best technology, combined with theory and craft.
4. The shifting paradigm of science is toward one guided by the coaction of science and technology, perceived as an integrated system.
5. There are speculations that an education in engineering may be the best education for a researcher in the natural sciences.
6. The success of team research is related to a proper mix of scientists and technologists, perceived as a cognitive intelligence system.
7. Technoscience offers new career opportunities for students, especially those with an interest in the computer sciences, mathematics, engineering, and material sciences, and any student with an "inventive mind," as well as broad multidisciplinary interests.
8. Science is a tool for generating new technologies, and technology is a means for extending the frontiers of science.
9. Computer networks are changing in fundamental aspects the way scientists work in collaborative research by providing a continuous flow of data exchanges.
10. Technology stimulates theory by making unplanned observations that then call for a new theory and interpretations.
11. Wherever possible, reference should be made to research problems that are now "on hold" for lack of appropriate technology to gather or process essential data.
12. To understand more fully the nature of contemporary science, every topic taught should conclude with statements on what we *don't* know about the topic, and where possible, why.

The transformation of our powers of observation and the technology for the management of large amounts of data emphasize that the practice of science comprises both theory and craft. Achievements are not the product of a "scientific method" and never were. We can expect more changes in the practice of science as the "information highway" develops and makes it possible to locate and access all the knowledge ever produced in the sciences.

CHAPTER 6

Biology in a New Vein

A widespread concern about education in the sciences at all educational levels, kindergarten through college, began in the 1970s. Much of the ongoing criticism centers on the failure of curricula to respond to the culture and practice of science as it is known today. Of the sciences traditionally taught in schools, biology has undergone the greatest transformations since the 1950s. What is now being sought are biology courses that reflect the role, practice, processes, and knowledge that reflect the realities of today's biology in contrast to traditional discipline-bound courses.

ORIGINS OF THE BIOSCIENCES

As with most new movements in a science, there is rarely a specific date or event that marks a turning point in the biosciences. It takes time to ripen a new point of view and even longer to assure its acceptance by a majority of researchers. To fashion a school curriculum that harmonizes with the practice and ethos of a new science usually takes decades.

The door toward a broader understanding of biology was opened in the late 1930s and early 1940s. The person most frequently cited as one who provided a stimulus for a new view of biology was Erwin Schrödinger, a Nobel laureate in physics. In a short book entitled *What Is Life?* (1944) he raised questions about the roles of physics and chemistry in explaining life processes. He postulated from his work in quantum mechanics that the atoms and subatomic particles that make up physical matter also account for the nature and behavior of living systems. He also noted that the "life cycle of an organism exhibits an admirable regulation and orderliness unrivaled by anything we meet in inanimate matter" (1944, p. 77). He acknowledged that the ways in which life works cannot be reduced to the ordinary laws of physics but will require a construction different from anything that occurs in a physics laboratory. Url Lanham views the beginning of today's biology in contrast to traditional views with the founding of genetics in 1900 (1968, p. vii).

What became common to biology, chemistry, and physics over time were methods by which micromolecules in living systems could be studied. The result was to shift biology from being primarily a descriptive science to being an experimental one at the molecular level.

TECHNOLOGY AND THE PROGRESS OF RESEARCH IN BIOLOGY

Optical microscopes have been the primary tool for observations in biological research since 1675. A fundamental change is now taking place as new instruments are invented. The scanning tunneling microscope with a laser attachment has made it possible to examine chemical bonding within living cells. The synchrotron enables researchers to examine the smallest units of life, the anatomy of biomolecules such as nanoparticles one one-thousandth of a micron in size. Other variations of modern microscopes include the X-ray, acoustic, photoelectric, and electron microscopes. The trend is to develop instruments that are not dependent on light absorption and emission. Each new type of microscope serves to expand the fields of biological research, sometimes by revealing unsought observations, and most of all, a view of life the likes of which we have never seen.

As you have been reflecting on sections of this chapter, millions of atoms have shifted and restructured their position in your brain, as Schrödinger speculated some 50 years ago. Microscopes and computers are now attempting to identify these movements and in the process are slowly converting psychology traditionally a social science into a study of biochemistry and biophysics.

WHAT IS MODERN BIOLOGY?

What's different about today's biology and its achievements cannot be identified by a precise statement. This revolutionary movement is sometimes described as post-Darwinian or post-traditional biology. Whatever name is used, it implies something different in kind from what has been taught in schools over the past 150 years. The current demand for curriculum reform in biology results from a public awareness that what is now in place in school biology courses does not reflect today's biology with its personal, social, and economic dimensions and its focus on human beings (Olson, 1986).

Today's biology has a host of different faces. Traditional disciplines (botany, zoology, physiology, anatomy) have been fragmented, hybridized, and reshaped in hundreds of new ways. We can glean the temper of these new fields of research from a sampling of their names: biophysics, global systems, biochemistry, biogeochemistry, genetic engineering, chemecology, biochemical engineering, bioinformatics, biogerontology, neuroscience, bioinorganic chemistry, molecular genetics, paleoecology, human engineering (ergonomics), biomedicine, organometallic chemistry, bioelectrochemistry, microbic ecology, computational biology, sociobiology, human gene therapy, bioethics, biomethanogenesis, biosensors, environmental biology, population biology, biotechnology, behavioral neuroscience, biomechanics, bionomics, neurochemistry, and biomedical engineering.

Many of these research fields interweave with the social sciences, such as economics, politics, and medicine. Biomedical studies are closing the long-standing gap between biology and the health fields.

The traditional discipline concept of "biology" has become a mosaic of life science fields. Each of these fields tends to fashion its own style of research, modes of inquiry, and ways of conceiving and interpreting data. The magnitude of change that forms today's biology has created an awareness that traditional biology is not rich enough in concept or method to answer the question "What is life?" or explain human behavior or the aging process of human beings.

The Life Sciences Come of Age

At the turn of the twentieth century the physical sciences dominated research in science. Physics was viewed as the "queen of the sciences." Today biology dominates the science research scene. Some idea of the extent may be gleaned from a recent analysis of the 100 most frequently cited research papers published during the period 1981–1988 (Holland, 1990). Of these papers, 96 were in various fields of biology and 4 were in physics. The neurosciences led in the number of citations listed, followed by studies in pharmacy/pharmacology.

For centuries, research in the sciences has focused on "exploring systems that are either intrinsically simple or that are capable of being analyzed into simple segments overlooking complex systems that just do not respond to an analysis of one factor at a time" (Ashby, 1963, p. 5). The new biology recognizes that human systems are interconnected and that the alteration of one factor acts as a cause to evoke responses in another or several other systems. For example, this is the way research in the neurosciences, biotechnology, and cognitive sciences operate. In modern biology, researchers are required to think differently from what is traditionally described in textbooks as "scientific inquiry" and "scientific method"; both concepts were modeled after traditional physics.

Research Practices in the Biosciences

The influence of technology on modes of research in biology was noted earlier. The problems faced in the new biology defy traditional methods of research. Problems today tend to be more holistic, more integrative, and more complex. New patterns of interpretation are to be expected when insights from several disciplines are combined, as in biophysics and biochemistry.

More and more, generating information on a problem is shaped by available technology; consequently, it is becoming increasingly difficult to separate a technological innovation from a biological achievement. This change led D. D. Price, a historian of science at Yale University, to define today's science as "applied technology" (1983). Bruno Latour describes contemporary research in science as "technoscience" (1987, p. 9).

Traditionally, significant research in biology leading to a discovery or a new theory was the product of one or two individuals and was identified by the scientist's name—for example, in the work of Louis Pasteur, Charles Darwin, William Harvey, Gregor Mendel, Jonas Salk, and James Watson and Francis Crick. Research in the new biology is done mostly by multidisciplinary teams. A problem involving micromolecules in brain actions will usually involve a research team consisting of genetic engineers, biophysicists, computer scientists, molecular biologists, biochemists, and psychologists. The team is viewed as a cognitive system structured to increase the fertility of ideas. For a problem that has proved to be extraordinarily difficult, for example, AIDS research, a number of teams representing research centers in different countries are organized into a collaborative effort using computers for instant communication between teams. In 1996, it was estimated that worldwide there were AIDS research teams in 50 countries involving at least 10,000 researchers. In addition, there are other research teams focused on the prevention of AIDS. These teams involve a mix of medical specialists, sociologists, social psychologists, and sometimes moralists, all working beyond the traditional laboratory-based experiment. In addition there are computer experts who specialize in the design of new approaches to AIDS research by using available data statistically likely to suggest where our ignorance about the problem seems to lie.

The most important intellectual skill associated with team research is the ability to communicate across fields. As the information superhighway develops, communication skills will become even more important as almost all the information ever developed on a problem can be accessed as needed.

The "pooling of minds" characterizes research in today's biology. Of the 12 most widely cited papers worldwide in science research for 1991, the number of authors per paper averaged 5.5 ("Vignettes," 1993, p. 180). No paper with a single author was recognized. The international record for a multiauthored paper as of now is 194 researchers for a study of respiration.

University-based scientists are the most likely to work alone, with only some help from graduate students. The reason is that university faculty members must establish a "paper trail" of published research to advance up the academic ladder. Team research is mostly identified with industry and the government, where 60% of all science researchers work today.

CONSTRUCTING ONE SCIENCE

From the very beginning of biology research in the United States, a distinction has been made in universities between "basic" research, focused on new discoveries likely to advance a discipline, and applied or practical research, with a concern for people and human progress. For 250 years universities have forced schools (by restricting admissions) to teach biology in the "basic" context, where the goal of research is to advance a discipline. To this day instructional goals and the selection of subject matter in biology have been in terms of acquiring laboratory skills, the learning of science processes ("scientific method"), with thinking skills embedded in a discipline (scientific inquiry) and a vocabulary of assorted technical terms found useful only by researchers. Rarely recognized is the fact that this approach makes school biology a career or occupational course.

Over the past half-century, biological research has moved into an integrative mode, represented by the examples listed earlier, such as biophysics, biochemistry, and biomedicine. This move has brought about revolutionary changes in how biological research is practiced. Notions of basic and applied research have become blended into a concept of "strategic" research that blurs such distinctions. Since 1945, around 40% of the Nobel awards have been for what in the past would have been considered applied research. The percentage may actually be even higher, depending on how one perceives applied or goal-oriented research, such as the work of Jonas Salk in searching for a polio vaccine.

Research in Biology Moves Toward a Human Context

Traditionally, the worthiness of research in biology has been judged by its contribution to advancing some aspect of a discipline, by theory, description, or experiment. Whether the research had meaning beyond the laboratory door or technical journal has traditionally not been a concern of the investigators. They value their work as "pure," "basic," or "fundamental," and at best, theoretical. The current shift to strategic research is a move from an esoteric view of biology to an exoteric one.

The transformation of biology more toward a study of human beings and their total environment includes both internal and external responses and their interrelationship. It is already apparent that humans do not respond to the realities of life in the same way as other organisms. For example, to understand human beings and their adaptive responses requires information from the social, cognitive, and behavioral sciences, plus more than a sprinkling of chemistry and physics. Research along these lines is illustrated in studies of the human genetic code and its implications for self-identity, health, and adaptation. Research in biotechnology is rapidly changing the world economy and laying the foundation for a new industrial revolution, as well as providing new insights into human behavior (Olson, 1986). Each of these examples illustrates the observation that research in contemporary biology is more socially than theory driven, more analytical than descriptive, and more focused on the human species than on lower organisms.

As biology has moved into an integrative stream harboring the national economy, social progress, human systems, and combinations of these factors, new realities have been spawned. In this explosive break with tradition, biologists now face questions of morals, ethics, values, and risks as their research becomes linked with human affairs. Take, for example, public issues raised about the human genome project. Should a human fetus be aborted if it is found to carry a disease-causing gene? Should geneticists be allowed to pursue a proposed global effort to survey genetic diversity among the world's peoples? The purpose is to seek new insights into our species— "from the prehistoric movements of peoples, cultures, and languages across the continents to the genetic basis of susceptibility to disease" (Kahn, 1994, pp. 720–722). The project would require DNA samples from diverse population groups from all over the world.

Public objections to the project are that it is "racist," that it fosters "genetic colonialism," and that it makes people "experimental animals." The U.S. government has stated that it will not support human embryo research, in such areas as human cloning, human–animal hybrids (the mixing of human and animal embryos), and "genetic diagnosis of gender in embryos that are expected to be implanted in women" (Burd, 1994, p. A32; Engelhardt & Callahan, 1978). While DNA fingerprinting is accepted by most of the scientific community for identifying individuals, it does not have the same status in the judiciary. The situation is true for a number of environmental issues, especially those related to the survival of an endangered animal or plant species. The cultures of biology and law each seek their own versions of what is to be considered truth.

This new dimension of biology where the public weighs the benefits of research has in turn led to a new field of study, that of bioethics. Courses on bioethics are widespread in schools of law and medicine, more so than in departments of biology.

Who Owns the New Knowledge in Biology?

The movement of biological research from university centers to industrial or corporate laboratories has raised questions about the ownership of new knowledge. Traditionally, new discoveries in biology are seen as belonging to anyone who wishes to make use of them. A common purpose of the 20,000 journals in biology and biology data banks is to pass on new information to other researchers. In contrast, biological research in industry is viewed mostly as providing a corporation with a product of economic value, such as new medicines, synthetic vitamins, vaccines, and improved varieties of plants and animals, mostly genetically altered. These are products for which a patent is sought, thus limiting what other researchers can do. Questions have been raised as to whether industrial researchers should have free access to research stored in data banks (Fletcher et al., 1994, pp. 75, 80). Is it unethical for a private firm to withhold information about a basic biological discovery until the time it becomes marketable in a new product? Cooperative endeavors between academia and industry complicate the issue, since research in universities is largely supported by the federal government.

Because it is biotechnology that is leading the emerging industrial revolution worldwide, research issues are being influenced by socioeconomic and sociopolitical consideration. We are now at the place in history where the natural sciences are called upon by the public to make contributions to society's well-being, the quality of life, and the common good (Carnegie Commission on Science, Technology, and Government, 1992b; Lewenstein, 1992; NAS, 1977; NSF, 1982). This view is long-standing. In 1620, Francis Bacon wrote, "The ideal of human service is the ultimate of scientific effort" and the purpose of schooling is "to equip the intellect for a better and more perfect use of human reason." Bacon proposed that the subject matter for education in the sciences be "that which has the most value for the welfare of man" (quoted in Dick, 1955, p. 441, 487). Alvin M. Weinberg has noted a critical issue in this regard as one in which it is scientists who provide the information for resolving a problem, but it is the citizen as a politician who decides how the information is used (Weinberg, 1972, pp. 209–222). He notes that often when citizens ask questions of scientists, though the information is available, bureaucrats determine the answers. In other words, the answer or decision transcends science. He coined the term "trans-science" to describe this situation.

NEXT STEPS

If biology education in schools and colleges is to harmónize with today's biology, it will be necessary to reinvent curricula from kindergarten through college. This challenge requires that supporting educational perspectives be derived from the practice and culture of biology as it exists today.

The American Association for the Advancement of Science in its report on Project 2061 stresses teaching science in the context of modern science, advances in related technologies, and the interaction of both in terms of shaping everyday life and the character of civilization (Project 2061, 1993). The growing connections between disciplines in the natural sciences and between the natural and social sciences are also recognized (Project 2061, 1990).

The National Academy of Sciences views education in the sciences from personal and social perspectives. The goal is to provide students with a vision and intellectual skills to use scientific knowledge in productive ways both personally and socially. The curricula designed for these purposes would also include the interrelationships of science and technology and their historical perspectives (NAS, 1994).

To effect modern biology curricula will require a mix of various kinds of biologists representing academia and industry, those specializing in societal goals, human behavior, and cognition. A mix of computer scientists is essential to deal with problems of accessing and communicating the ever new cascade of knowledge in the life sciences. Scholars from the humanities are needed to facilitate the integrative themes between the natural and social sciences in terms of human aspirations and welfare. Today's biology is to be viewed as a way of life, not simply a body of information.

To keep biology curriculums up to date, the federal government has a responsibility for ongoing research related to the sciences and for publications on research achievements in the sciences selected for their relevance to school life science courses. These reports should be written in civic language so as to serve not only schools, but also the interested public. The advancement of a science/technology–oriented democracy rests upon a continuing program of intellectual nourishment, especially in a knowledge-intensive society.

CONCLUSION

The development of today's biology has its roots in the work of Gregor Mendel, later augmented by the Watson-Crick model of DNA structure. Schrödinger's reflections on the molecular basis of life redirected much of biology from purely a matter of observation and classification to an experimental science. There has also been a change to a greater concern for the human species.

As the twentieth century comes to a close, we are entering an age of biotechnology, that both influences the global economy and benefits human welfare. There is no doubt we have arrived at the door of a golden age for research and study in biology. Today, biology is viewed as a social investment in terms of its benefits to individuals and to society.

Now the problem is to bring school biology curricula into harmony with the image of today's biology. So great are the distinctions between the old biology and new biology that simply trying to revise traditional courses will not resolve the problem of modernizing school biology courses. The next task is to define the purposes of an education in biology for all citizens. The perspective is one of helping human beings live more sensible lives and play more responsible roles as citizens (Hurd, 1971a). Traditional biology courses have been designed to prepare students to understand the structure of a discipline as preparation for an academic career in biology, not the reality of biology in today's world and in their lives. In this context, the need is to reinvent biology curricula. The approach is in contrast to the traditional practice of simply revising and updating biology subject matter for its own sake, from time to time.

CHAPTER 7

The Developing Adolescent: An Endangered Species

As our nation moves into a knowledge-intensive era, it offers a wealth of promises for the welfare of today's developing adolescents, such as more career opportunities and a greater diversity of roles as citizens. The purpose of this chapter is to provide a snapshot of the problems and issues early adolescents are experiencing in our culture, and how they are coping and adapting. Needless to say, developing adolescents are having trouble as they try to find their way through the maze of family, biological, and social changes with little help in or out of school. Early adolescence is viewed by too many adults as an incurable disease, much like the common cold (Amour, 1963, p. 19).

As we shall explore later, problems of adolescent development have been increasing for over 100 years: first, at the turn of this century, when the nation moved from an agricultural society to an industrial one, and second, now, with our transition from an industrial era to a knowledge-intensive one. In both instances adolescent attempts to cope and adapt to an evolving culture and social changes have been characterized by a variety of high-risk behaviors such as violence, illicit drug and alcohol use, premature sex, gang life, and negative stress behaviors.

Only recently has adolescent development been the subject of serious research by biologists, sociologists, psychologists, psychiatrists, and other medical specialists. There is now a growing body of findings that include elements of cognitive social, biological, and emotional development. What has not yet been considered are the implications of these findings for the development of science curricula in the middle grades. What is required is a science curriculum in the context of the early adolescent and the total cultural scene of life, rather than the structure of science and its traditional modes of inquiry. This is not to say that all the problems of adolescent development are the responsibility of schools. Parents and communities also have a responsibility.

THE ROOTS OF EARLY ADOLESCENT PROBLEMS

An awareness of problems of early adolescent development emerged a century ago, in the 1890s. At that time the nation was moving from an agricultural society to an industrial one. The trend caused families to move from farms into cities, where they met social and economic conditions for which they were not prepared. Least prepared among them were emerging adolescents.

The problem for education was first recognized by Charles W. Eliot, president of Harvard University and chairman of the National Education Association Committee of Ten. To deal with the adolescent problems, the Committee of Ten, reporting in 1894, recommended that schools be reorganized with six grades for the elementary school and six upper grades, thus placing developing adolescents with more mature adolescents (Glatthorn, 1987). The rationale for this arrangement was that young students would thus be better prepared later for entrance into a university. The welfare of the early adolescent as a young person with personal and social adaptive problems was not mentioned. The 6–6 grade organization of schools was never implemented.

As might be expected, the problems of the early adolescent became more severe. The use of illicit drugs and alcohol increased, as did violence. These risk-taking behaviors represent the way a troubled or confused adolescent seeks to meet the challenges of a changing society. A first attempt to study the plight of adolescents was done by G. Stanley Hall (1905), a psychologist who wrote two books on the subject. There is no record that his recommendations influenced changes in schooling in any way.

In 1909, it was again recognized that something should be done in schools to improve adolescent welfare. It was recommended that the school systems be reorganized into a seventh-, eighth-, and ninth-grade unit, the junior high school. This arrangement would serve to isolate developing adolescents from both younger children and the influence of more mature adolescents. Specially trained teachers and curricula designed "to meet the needs of students" were also recommended. But the curricula were never developed. Science courses in the junior high school became mirror images of high school courses, in both their goals and their subject matter.

In 1915, a new science course was developed and named "general science." It was largely a study of the practical or technological applications of science. The physical science part of the course explored the internal combustion engine and the automobile, the airplane and how it flies, the radio, electrical appliances, and safety precautions. Biology emphasized conservation and the environment, some human anatomy and physiology, and the evolution of plants and animals. The educational rationale for the course was to elicit student interest in taking more science courses in high school.

General science became a popular course with students but failed to generate in them an interest in taking additional science courses in high school.

Following World War II, the United States moved into new fronts in both science and technology, characterized as an air age, then an atomic age, and later an electronic age. These changes created a demand for more scientists. In the early 1950s, the newly organized National Science Foundation (NSF) was formed by the U.S. government to oversee the task of enlisting more students into science careers.

In the 1960s, NSF supported a national course content improvement program in school science and mathematics. Scientists were asked to take charge of the reform effort, and they chose to revise science courses to emphasize the structure of science and its modes of inquiry. Middle grades science programs were made more academic and rigorous in order to better prepare students for more advanced courses in high school (Hurd, 1969a, 1970a).

During the 1960s and 1970s an awareness developed again that early adolescents were having more and more difficulties growing up and meeting the adaptive demands of a rapidly changing society. The first proposal then for meeting the needs of early adolescents was to change the organization of school grades by combining grades 6, 7, and 8 into a middle school, replacing the junior high school arrangement (Hechinger, 1992). A total of 1,662 middle schools were formed by 1970 and 4,329 by 1989. A total of 4,711 junior high schools existed by 1970, and 2,181 by 1989. However, the majority of students continued to be enrolled in the traditional kindergarten-through-grade 8 schools.

A National Middle School Association (NMSA) was organized in the late 1960s to revitalize the education of young adolescents. The document *This We Believe* (NMSA, 1995) was developed by middle school teachers and professional educators to outline a rationale for a middle school education that recognizes the "diverse characteristics and needs" of students between the ages of 10 and 15 years (1995, p. 5). It was recognized that adapting to modern life in a changing society and a new world of science and technology demanded a curriculum responsive to these conditions. A vision of this curriculum is described in terms of characteristics of the early adolescent by the National Middle School Association (1995, pp. 35–40).

Although science curriculum goals for the middle school were fashioned in the 1980s, the supporting curriculum has failed to materialize. Science textbooks, course instruction, and tests do not differ from those of the junior high school (Lounsbury & Clark, 1990). Middle schools have responded to reform efforts by changing the lengths of class periods and the school year. Team teaching is used more widely, and teacher certification has changed (August, 1989; Hurd, 1984). The concentration of reform efforts has been on governance, management, and the operation of schools, not on a curriculum tailored to meet the needs of early adolescents in a changing society and culture.

A Generation at Risk

The host of changes taking place in our culture and in society over the past century have resulted in a generation of early adolescents who are increasingly at risk. Neither teachers and educators nor parents are aware of just how to deal with these problems. It is obvious that changing combinations of grade levels in the school system does not work. Making science and other courses more rigorous does not work, either. One gets the impression that too many adults view early adolescence and its problems as a natural period of turmoil in human development and seem to think that if adolescents are left "to do their own thing" they will eventually get over their troubles and confusion.

A full issue of *Daedalus* published in 1971 opens with the statement that "We know less about early adolescents than we ought to" (1971, p. v). Over the next 25 years a massive amount of research and study was focused on problems and issues of early adolescent development. Commonly used phrases to picture the condition of today's 10- to 15-year olds are "an endangered species," "an imperiled generation," "growing up forgotten," "caught in the middle," "misguided and at risk," "roleless in today's society," "a human tragedy," "products of an education wasteland," and "victims of an educational scam."

Any serious considerations for reforming middle grades science education to meet the needs of early adolescents must begin with a study of the mismatch between schooling and how 12- to 15-year-olds live their lives (Beane, 1990; Hamburg, 1987; Lipsitz, 1977; Lounsbury, 1991). Currently, efforts to establish national educational standards do not reflect this situation, nor has this traditionally been the practice (Lounsbury & Clark, 1990).

Today's early adolescent is not the same kind of individual as the early adolescent in the 1950s and 1960s. Social, cultural, economic, and biological factors currently influencing adolescents are different from what they were in past generations (Feldman & Elliott, 1990). Since 1970, family structures have changed dramatically (Fuchs & Reklis, 1992; Office of Educational Research and Improvement, 1991; Wetzel, 1989). Beginning in the 1960s, poverty and fragmentation of the family had far-reaching effects on the physical, mental, and emotional well-being of young adolescents. The trend for grandparents to move from family homes to distant retirement communities has taken away the one dependable source of sympathy and a shoulder that young adolescents could traditionally cry on. Middle grades teachers are finding that they are increasingly drifting into the role of surrogate parents.

Essential to determining a viable curriculum for early adolescents is a clear picture of their developmental issues. There is a consensus that these tasks include: (a) emancipation from parental attachments; (b) development of satisfying and self-realizing peer attachments, with the ability to love and appreciate the worth of others as well as oneself; (c) an endurable and sustaining sense of identity in the familial, social, sexual, and work-creative areas; and (d) a flexible set of hopes and life goals for the future (Berkovitz & Sugar, 1986, p. 3; Report of the Panel on Youth, 1974).

Early adolescents today are not as healthy as is usually taken for granted (American Medical Association, 1986). The prevalence of drug use, stress factors, increase in suicide rates, early pregnancies, and poverty all point to the fact that growing up today is harder than in past generations (Hechinger, 1992; Hendee, 1991; Institute of Medicine, 1993; Millstein, Peterson, & Nightingale, 1993; Panel on High-Risk Youth, 1993; U. S. Congress, Office of Technology Assessment, 1991; U.S. Department of Education, 1991).

Symptoms of Early Adolescent Maladaptive Behavior

Following is a summary of troubling issues reflected in the life conditions and behavior of early adolescents. It should be recognized that these symptoms reflect serious social and educational problems. The list of risk behaviors provides a knowledge base for deliberations on the reinvention of science curricula as well as a view for public policies on the well-being of youths. The data presented are limited to recent surveys of early adolescents. Longitudinal studies of the transition of early adolescents to adult life are lacking. It is safe to say that no single student has all the problems listed, or that all students are victims of risk-taking behaviors. It is estimated by those who have studied the problems of the 10- to 15-year-olds that around 50% experience one of these problems, such as depression, suicide, loss of self-esteem, or unnecessary ill health. The problems also vary with culture, ethnic background, and a variety of family conditions. It is not assumed that schools can deal with all of the problems.

A criticism of the current education reform movement in education is that we have forgotten the students in the middle grades, failing to recognize that they are not like either children in elementary grades or students in high school. Efforts over the past 100 years to isolate the developing adolescent in special grade levels has not resolved developmental problems. The concerns identified by Hall in 1905 persist today. Following is a listing of the symptoms that represent maladaptive behaviors in adolescents growing up culled from studies ranging from 1979–1995.

- Drinking and use of illegal drugs. Of those who drink in high school, 56% began in grade 6.
- Illegal drugs have been tried by 51% of students by grade 7.
- Deaths from suicide are increasing, and are in 1996 at an average of one every 90 minutes. Motor vehicles lead as the cause of death for ages 10 to 14 years. The homicide rate for early adolescents doubled between 1985 and 1992.
- The age of onset of puberty for girls 150 years ago was 16 years; today it is 12.5 years. By age 16, 30% of early adolescents are sexually active. Of this number, 23% have acquired a sexually transmitted disease, usually gonorrhea.
- School dropouts begin in the middle grade and increase in number with each passing year. In our knowledge-intensive era, early school dropout represents a fatal economic disease, stemming from ignorance. School dropout rates vary with gender, ethnic group, and family structure. Six percent of all students now drop out before grade 10.
- A crucial factor in adolescent development has been the changes in family structure. Thirty-one percent of American families have only one parent at home, and the percentage of single-parent households is increasing. Of this number, 27% are headed by a mother and 4% by a father. Five percent of families include a grandparent. Forty percent of adolescents say parents are unavailable to them; TV is the substitute. Only 6% of developing adolescents now live in a "traditional" American family consisting of a working father, a mother at home full-time, and two children.
- Fifteen percent of early adolescents belong to gangs, often viewed by them as family.
- Health, biological and behavioral, is not as good as is generally assumed. Professional health care is not affordable or not deemed necessary by a majority of families with early adolescents.
- An increasing percentage of early adolescents live at or below the poverty level; 25% are white and 66% are black, and the percent is increasing for both groups.
- Forty-seven percent of early adolescents have at least one of the problems listed above, and 20% have two or more.
- Violence is increasing: 40% of early adolescents know someone who was shot to death; 135,000 students bring guns to school; 77% of city schools have police patrolling the grounds; 19% of schools have metal detectors at school entrances.
- In an average week, 27% watch at least 21 hours of television; 30% spend an average of 2 hours per day alone; average time spent interacting with parents is 20 minutes with a mother and five minutes with a father. Six hours per week per teenager are spent talking on the phone; 1.8 hours per week are spent reading for pleasure; 5 to 6 hours per week are spent

on school homework; 2.8 hours per week are spent "hanging out" at the mall, and the same number of hours are spent shopping for oneself. One-third of eighth graders read fewer than 5 pages per day; 5 hours per week is the average spent playing sports.

Primary sources of information on the study of the early adolescent are American Medical Association, 1986; California State Department of Education, 1987; Carnegie Council on Adolescent Development, 1989, 1995; Center for Early Adolescence, 1988; Center for the Future of Children, 1995a; Children's Defense Fund, 1988; Collins, 1984; Grant, 1988; Hurd, 1978b; Katzenmeyer, 1979; Lewis, 1993; National Research Council, 1993; Panel on High-Risk Youth, 1993; "Adolescence in the 1990s," 1993.

The problem attracting the most attention in adolescent development is that of violence. It is only recently that violence has become a matter of scientific and social research to determine an understanding of and strategies for prevention and control (Panel on Understanding and Preventing Violence, 1993). To fully understand problems of adolescent development, it is necessary to recognize that adolescents are both biological and social beings. Feldman and Elliott (1990, p. 7) have noted that "biological drives and social prescriptions combine to define adolescents in the United States today." Violence is a symptom of maladaptive behavior brought by the host of changes taking place in our society and culture. How to resolve the problem is the subject of much debate and some scientific study (American Psychological Association, 1993; Hamburg, 1993; Johnson & Johnson, 1995; National Issues Forums, 1994).

How do young adolescents view their problems? Their answer is that "nobody seems to care" and schools "don't teach you anything about yourself or other kids" (Metropolitan Life Survey, 1996).

THE NATIONAL CURRICULUM PROJECTS

The American Association for the Advancement of Science (AAAS) has developed a group of science curriculum benchmarks entitled *Science for All Americans* (Project 2061, 1990). The authors outlined a science program deemed suitable for early adolescents. The major topics include "human identity, human development, the basic functions of the body, learning, [and] physical and mental health" (AAAS, 1990, p. 71). The project includes perspectives on human behavior derived "from the findings of the separate disciplines within the social sciences—such as anthropology, economics, political science, sociology, and psychology—but without attempting to describe the findings themselves or the underlying methodologies" (p. 87).

The National Research Council (NRC) has developed National Science

Education Standards for kindergarten through grade 12. Standards for grades 5 to 8 include a dimension in the life sciences. The recommended standards are "from the point of view of individual organisms to recognizing patterns in ecosystems and developing understandings about the cellular dimensions of living systems" (NRC, 1996, p. 155). Personal and social perspectives on science are mentioned but not defined in terms of adolescent development or welfare.

The National Science Teachers Association (NSTA) curriculum project *Scope, Sequence, and Coordination* specifically targets middle level students for the purpose of generating in them an interest in science during their formative years and in encouraging them to continue studying science (NSTA Committee on Scope, Sequence, and Coordination, 1992, p. 15).

Each of these education reform projects identifies the nature and structure of traditional science disciplines as the appropriate context for the teaching of science in the middle grades. What is not recognized is that adolescents and scientists live in different worlds and perceive problems of life and living in different contexts.

The basic questions to be asked of these projects have been under debate for some 140 years. In 1847, the issue of "science for all" was raised by James J. G. Wilkinson, a member of the Royal College of Surgeons of London. He saw the issue of education in the sciences as one in which "the end for which knowledge was sought and recorded by the learned, and the end for which it is required by the multitude, are not the same, but different ends" (Wilkinson, 1847, p. 3). By this statement he meant that science standards for public welfare and for career preparation are different. He also pointed out that the reason science is so easily forgotten by most students is that the subject matter is not perceived by them as serving a useful purpose in their lives.

Herbert Spencer in his famous 1859 essay "What knowledge is of most worth?" states the issue as follows: "the worth of any particular order of information does so by showing its bearing upon some part of life." Spencer saw the essential question for schooling in the sciences as the "effects which these facts can produce on human welfare." He saw the importance of knowledge in terms of its usefulness to personal well-being. "How to live?—that is the essential question for us. Not how to live in the mere material sense only, but in the widest sense" (1859, p. 14).

While policy statements for studying the human species are recognized by Project 2061 and the NAS standards, the subject matter is not specifically identified. Science literacy and scientific inquiry are viewed as major goals for the teaching of science; however, both of these are in the context of academic science. The benchmarks and standards described in these curriculum reform projects reflect the opinions of academic scientists, neglecting the opinions of strategic scientists. Yet it is strategic scientists who carry out nearly 75% of all scientific research today. The ways in which they practice science are more attuned to the way in which citizens use science in resolving problems and issues. For public understanding of science, the goals are social inquiry, cultural literacy, and skills in the wise utilization of science findings.

The Task Ahead: Meeting Adolescent Needs

Science curricula sensitive to meeting "the needs of developing adolescents" have yet to emerge. Whenever the issue has arisen, curriculum developers have rooted their recommendations for changes in terms of updating only the subject matter to be taught. There has been a failure to recognize that early adolescence is a period of social, biological, and intellectual changes.

For centuries the traditional illusion of science curriculum reform has been attempts to expand concepts and methods of science as known in the laboratory. This approach is in contrast to a science curriculum created in the context of students for life in the real world, the total scene of life. Furthermore, a science curriculum is sought that is oriented to the future rather than merely a review of the past, justified as basic. The U.S. Congress has been briefed on the need to invent new curricula in terms of the student (Beane & Lipka, 1986; Furby & Beyth-Maron, 1990; Hamburg, B. A., 1990; Hamburg, D. A., 1995; Hurd, 1995; Morganett, 1990).

What should be a framework for a middle grades science curriculum has been described in terms of life skills (Furby & Beyth-Maron, 1990; Hamburg, B. A., 1990; Morganett, 1990; NW Education, 1996). Many of these skills are linked to practical thinking, problem solving, and making judgments and decisions, with considerations of values, ethics, and the common good (Hurd, 1989a). A criterion of a life skill is that it can be applied in daily living and in ways apparent to the adolescent:

> Contemporary adolescents need help in acquiring a range of social competencies to cope with academics, to meet fundamental challenges of forming stable human relationships, to maintain hope about their future, to understand and adopt health promoting behaviors, to make wise decisions about life options, and to optimize use of social networks. (Hamburg, B. A., 1990, p. 3)

Middle Grades Reform: A False Start

During the current educational reform movement that began in the 1970s, a number of middle school programs have been identified as excellent, exemplary, successful, or innovative (Hurd, 1981; Lipsitz, 1983). After reviewing the reports and visiting some of these schools, I found them to have a distinctive school culture. There was no paper on the floor and no graffiti on the walls inside or outside the school. Students had improved their grade averages, school attendance had improved, scores were up on standardized tests, more parents were involved in school activities, and most teachers were deeply concerned about their students. Students stated that they liked their schools because "teachers call you by your first name" and there are lots of social activities.

These exemplary schools demonstrated that a favorable climate for learning can be created, and research shows this to be a significant factor in student achievement, whatever the educational goals (Rutter et al., 1979). However, the science curricula in these schools were no different from those found in other schools. The textbooks used were the same, and classes were taught and students were tested in the same way (Lounsbury & Clark, 1990).

In the 1980s and 1990s there have been a host of unfocused reports and recommendations characterizing early adolescents, the qualifications of their teachers, and international comparisons of student achievements. Schools have responded by changing the length of class periods, establishing longer school years, and buying more computers. New standards have been proposed for teacher certification and for professional development (August, 1989; Hurd, 1984). In many schools, reform efforts have consisted simply of educators working harder and longer on their present curricula, not recognizing those curricula are the source of the crisis in science teaching. The concentration of reform efforts has been on governance, management, and operation of schools, not on curricula.

TOWARD A NEW VIEW OF SCIENCE CURRICULA

We have seen the problems of today's adolescent, and some aspects of the changing world in which they must live, such as the information age, a global economy, and new ways of work. At the same time, the nature and ethos of science is changing. We are living in a new era of civilization to which everyone needs to adapt if they are to experience an acceptable quality of life.

America 2000: An Education Strategy (Alexander, 1991) describes the task ahead for the reform of schooling as one "that sets aside traditional conceptions" and makes "revolutionary changes." The curriculum endeavor should represent a "quantum leap forward" with goals that are "bold and long-range."

Attempts to revise, refine, reorganize, restructure, revitalize, and make today's science curricula more rigorous ends up causing matters to worsen. What is sought are science courses focused on what adolescents feel is worth knowing and is relevant to daily living and a changing society. In other words, this would be a science curriculum in the context of the lived world to which adolescents must adapt, and in a few years a world they can help to design.

The New World of Work

Our information-based society has brought about revolutionary changes in the world of work. Over the past 100 years of our industrial society, the majority of jobs depended upon a worker's learning one or two mechanical skills. These were acquired on the job or in school vocational courses (e.g., typing, shorthand, bricklaying, carpentry, welding) or in manual training high schools. The workforce skills for an information age center on intellectual power such as learning to learn, higher order thinking skills, problem solving, informed decision making, and responsible judgments. Connecting schooling to work has always been an educational problem with few attempts made to resolve it (Wirth, 1992). In 1994, the U.S. Congress passed the School to Work Opportunities Act (STWOA) to foster the preparation of young people for the requirements of a workforce in which they will spend their lives. Here there is a role for both schooling and the workplace; the reform task is one of blurring the distinction between schooling and working (Hurd, 1994b, pp. 103–116).

The economics of the real world today is one that views every individual as "human capital" or a "human resource." The intellectual skills and insights required for economic success in the new world overlap those now accepted in desirable purposes for the teaching of science (Hurd, 1989b; Marshall & Tucker, 1992). The Office of Technology Assessment of the U.S. Congress has recommended that school programs related to learning to work should begin no later than the seventh grade (Office of Technology Assessment, 1995, p. 3). Young adolescents are vaguely aware that career opportunities are changing, but not in ways adolescents understand. The current wave of "downsizing" in the workforce involves workers who have failed to keep up their learning or who can be replaced by robots, such as the automatic tellers in banks. The path of economic success in life is paved with knowledge. The time has arrived to close the gap between schooling and working (Wirth, 1992).

CONCLUSION

This chapter about the developing adolescent in today's society and in schooling seeks to portray the maze of conditions that call for a reform of education in the sciences. None of us has had experience in what it means to move into a new culture. Neither have we had experience in designing an education to meet the demands of life in the near future and beyond the year 2000. For over 200 years, and unfortunately, continuing, science education has been rooted in the past and reforms have been restricted only to updating subject matter to improve an understanding of a discipline. Today, the call is for a reinvention of science curricula that will have meaning for individual welfare, social progress, and economic advancement in a knowledge-based global economy. The curriculum sought is one based in the problems and issues that developing adolescents endure in our changing culture.

CHAPTER 8

Science Education
for a Changing World

This chapter sketches the implications of the earlier chapters of this book for the reform of science teaching at all educational levels. At no period in human history have so many interacting changes taken place in a society in so short a period of time, each calling for a new view of education in the sciences. Educational responses to these demands have been under way since the early 1970s, but with few identifiable results except confusion. Progress has been sidetracked by a maze of blind-alley approaches, such as lengthening the school year, requiring more science courses for high school graduation, making science courses more rigorous, demanding higher test scores on outdated tests, and comparing these scores with those of students in foreign countries, or simply modifying existing science courses by updating or revising them, and other sorts of tinkering. Reform efforts are for the most part trapped in tradition.

A first step in designing a modern school science program is to identify the reasons or elements that frame the demands for a new curriculum. We are now living in a science/technology–oriented democracy. At the same time, the nation has moved from an industrial-era culture to a knowledge-intensive society. These shifts have influenced the world of work and the requirements of citizens for participating in the public economic and social life of the nation. In this context each of us is viewed as "human capital" and the nation's major resource for shaping its future. Preparing students for the future, the world they will live in, requires that more emphasis be placed on the cognitive functions they will need in order to seek, access, and organize information in ways that will be productive in resolving the changing problems of life and living.

A major social change that has taken place over the last half-century is the structure of American families and the lives of children and adolescents. We have before us the task of providing a valid education in the sciences that increases the adaptive capacities of students biologically, socially, and economically. Changes in the culture and practice of science have outmoded the traditional goals and curricula of school science. These conditions and others call for a new vision of education in the sciences and then the invention of supporting curricula. The remainder of this chapter will focus on frontier issues for the modernization of an education in the sciences.

MODERNIZING SCIENCE CURRICULA

What factors are necessary to consider in framing an education in the sciences for the year 2000 and beyond? We need to recognize the current changes in the nature and practice of the natural sciences. Over the past century the concept of science disciplines has lost its meaning and is being replaced by thousands of research fields and many styles of research. At the turn of the twentieth century, the "queen" science was physics. Today, biology is gradually absorbing the physical sciences and generating new fields such as biophysics, biochemistry, bioengineering, and biogeochemistry. Since the 1500s the function of the natural sciences has been seen as basic research emphasizing theory and universal principles such as evolution, relativity, and quantum theory. Today the major emphasis is upon strategic or mission-oriented research, including predetermined social, economic, or human ends. Modern science curricula thus need to include a dimension that is focused on the utilization of science knowledge for resolving personal-social and civic problems as well as fostering social progress.

Traditionally, science courses have focused on skills deemed essential for scientific inquiry and a career in science. This career-oriented goal for every student can no longer be justified for life in a knowledge-intensive society. Life skills are now sought by which knowledge from the sciences can be utilized for improving the quality of life for all students throughout their lives. Life skills are in harmony with the purposes of strategic research in the natural sciences. Another clue for reinventing science curricula is found in the ongoing integration of the natural and social sciences, along with aspects of the humanities. For example, "human behaviors are found to be biopsychosocial in nature" (Frankel & Teich, 1974, p. 125).

AN EVOLVING HUMAN SCIENCE

This chapter proposes to identify a philosophical base to guide the reinvention of science courses. Without a theoretical base, there is no way to judge the validity or quality of a proposed science curriculum for a new age of both science and society. For the past century every science curriculum reform movement has included the goal of "meeting the needs of students."

What this purpose should mean for the welfare of the student has not been identified. There is an unwritten contract between schools and the government that schools have the responsibility for preparing young people for life in a democracy. There is also an obligation to prepare students for self-understanding and achieving a high quality of life. To approach these ends requires a new order of education in the sciences.

The science curricula we need are those that will help close the gaps between science and human affairs. Such programs should emphasize the utilization of science knowledge to extend human capacities as a social as well as a biological species. This means a focus on the operational use of the findings of science to enhance human adaptation. By contrast, the traditional view is that a science course should stress the structure of a discipline from an academic and a career viewpoint, a view that sees a school science as preparation for college rather than preparation for life.

A position to begin with for the reinvention of school science courses is to recognize children and adolescents as human beings. This stand recognizes that as a human species we have the power to control and direct our own development and evolution, a process that requires unending learning. We also need to recognize that *Homo sapiens* is a social animal, the product of both biological and cultural evolution. To understand oneself requires an integration of information from the fields of anthropology (cultural and physical), geography, sociology, economics, and the strategic sciences.

A number of courses in human biology have been developed since the 1930s, but most have explored only the biological characteristics of human beings, much like a frog is studied. The proposed human science is a more comprehensive view of ourselves; it seeks to link human beings with their culture and with modern science and technology, each in a biosocial context characteristic of human ecology.

It is within the framework of a human science that we find a base for "meeting the needs of students." "Towards a Human Science" was the title of Margaret Mead's talk as president of the American Association for the Advancement of Science (Mead, 1976).

Productive Knowledge

During the past 25 years of the current reform movement an issue that stands out is that student science achievement is no higher today than in the past using traditional curricula. The seriousness of the problem is augmented by the nation's movement into a knowledge-intensive age, where knowledge is seen as our most important natural resource. For the individual, productive knowledge is the basis of one's human capital. Educators and the public press uniformly describe the national issue as the need "to create a nation of learners." Knowledge and its utilization have become a lifetime career for all students if they are to be economically successful. The Organization for Economic Co-Operation and Development (OECD) involves 26 developed countries, including the United States. At the 1996 meeting of OECD, all countries "accepted lifelong learning for all as the guiding principle for policy strategies that will respond directly to the need to improve the capacity of individuals, families, workplaces and communities continuously to adapt and renew." Furthermore, active learning was viewed as essential "in promoting economic development, democracy and social cohesion in the years ahead" (OECD, 1996, p. 13).

Since our founding, school science has been taught in the context of America as a producer of new knowledge about the natural world. Today the focus changes to developing learning skills that are essential for a consumer who utilizes science information. This means the subject matter of science is selected for its value in resolving personal, social, and economic problems, and for augmenting our adaptive capacities as human beings. This movement is recognized in such educational catch phrases as the "socialization" and "humanization" of science to improve "the public understanding of science." The nature of the science knowledge to be taught in a six-year core program required of all students is yet to be validated and synthesized. The process will result in a presentation of science in a personal-social context.

A Lived Curriculum

A lived curriculum is a science of ourselves. It focuses on our understanding of what it means to be a human being as a biological and social mammal. The purpose is to link the natural sciences and technology to the personal and social development of adolescents in ways that enable them to use the information for decision making in a science/technology–oriented democracy.

The roots of the lived curriculum are in harmony with the new forms of strategic research that characterize the nature and practice of contemporary science, where knowledge is sought for the benefit of society and humanity. This curriculum is also anchored in the cognitive sciences, where the nature of learning is being reexamined. It is becoming clear that science knowledge has a longer tenure in memory when students can experience it and recognize that it can be used to help them understand themselves and can be used in daily living. Confucius, 2300 years ago, recognized this principle with his comment that "knowledge is made to use." We also need to recognize that knowledge has an organic quality, expanding in meaning or significance each time it is used. This perspective is in contrast with the traditional science knowledge that is isolated in separate disciplines with little meaning beyond the classroom door, a position now seen as putting knowledge in an intellectual straitjacket as far as life experiences are concerned.

The science courses now being sought are those developed in the context of students and their lived-in world. The purpose is to provide a curriculum students can experience in their personal and social affairs and can use for making rational decisions that serve human adaptive capacities. The conceptual issue is a curriculum concerned with what makes human beings human.

The themes for a lived curriculum are drawn from "real life" and the "real world." The science/technology base comes from the unification of a wide range of contemporary research fields, including anthropology, psychology (individual and social), neurobiology, sociology, cognition, human ecology, behavioral sciences, geography, genetics, biotechnology, economics, medical sciences, and many more (Manen, 1990). In a lived curriculum, typical themes represent the functional or adaptive needs of students rather than discipline structures (Bandura, 1977; Becker, 1971; Cohen, 1968a, b; Corney, 1972; Doby, Boskoff, & Pendleton, 1973; Environmental Studies Division, Office of Research and Monitoring, 1973: Gibson & Ingold, 1994; Hardin, 1959; Harrison, 1993; Ray & Nelson, 1971; Trigg, 1985; Wallace & Srb, 1961).

The following themes were randomly selected from those now being developed in schools and colleges. The ones most commonly listed were related to health (biological, behavioral, social), including wellness, fitness and safety; environment (human and global ecology); science and technology (personal, social, and cultural interactions); factors in learning how to learn and process information; biotechnology (genetic engineering); humans at risk (drugs, alcohol, tobacco, violence, and mental depression); optimizing the functions of human biological systems; decision making (individual and cooperative); understanding other people; adapting to change; strategic research in the service of human and social affairs; life skills (personal and social); knowing oneself; resolving real life problems; quality of life; and many more. The tone of these themes is a science that students can care about and live. It should be noted that most themes blur traditional disciplinary lines.

Adventures in Learning: Lived Investigations

Teaching science in a personal-social context brings into question the place of laboratory and hands-on activities. Traditionally, these activities have been viewed as important for acquiring intellectual and mechanical skills for the practice of science. Students perform experiments, and if directions are followed, everyone will get the same answer. Children in the early grades are engaged in "hands-on" activities which are expected to convey "the joy of discovery." Children do say the activities are fun, but rarely can they verbalize the science concept embedded in the activity. A weakness of "hands-on" activities is that they are designed by adults as they themselves perceive the desired concepts. Louis Pasteur in the 1850s noted that children will acquire the intended science concept only to the extent that they have been prepared to do so. Science concepts rarely emerge by a student's simply handling objects in different ways; concepts are in the mind and not in the fingers.

A central purpose of the science education reform movement is the public understanding of science. The goal is for students to know science in ways that let its findings be effectively used in personal-social, social-civic, and economic relationships (Thayer, 1938). This purpose is contrary to the ways science is traditionally practiced in the laboratory. The new perspectives of science education are reflected in the nature and practice of strategic research in the sciences. The National Science Foundation (NSF) sees the issue this way: "in a democratic society that is highly dependent on science, mathematics, and engineering, the scientific enterprise cannot thrive unless it is open to all segments of the population" (NSF, 1995, p. 18).

To involve students in the variety of ways that science/technology is influencing their adaptive capacities and their lived-in world requires new procedures for teaching and learning in science. One change is a vision of the science classroom as a learning center where class and laboratory activities are combined in a program of investigations. Investigations directly involve the students as a part of the research. For example, a morning newspaper features the ethical problems associated with the use of animal parts such as a pig's heart to substitute for a failing human heart. Here the student takes a position to investigate. Investigations do not follow a formula typical of school laboratory experiments; they are more a matter of finding relationships, sequences, regularities, connections, and meaning. The findings are more correlational than experimental.

Problems related to science in a personal-social context require ways of thinking that are different from those required for scientific inquiry. Investigative problems have more variables that cannot be controlled, and findings are more difficult to interpret (Aikenhead, 1985). Schools are now pressured to teach higher order thinking so that students have a better chance of resolving their own problems, engaging in the nation's affairs, and competing economically in a knowledge-intensive society. The reform movement centers on the development of life skills rather than laboratory skills. The life skills most sought are those related to decision making in such contexts as making judgments in different frameworks like law, ethics, policy, values, morals, the future, and yesterday's findings in relevant fields of science.

Investigations tend to be future oriented in comparison with the earlier use of experiments to verify past achievements in the sciences.

Investigations involving human affairs as well as strategic research in the sciences typically require a multidisciplinary approach, a linking of the natural and social sciences. This practice in strategic science changes the philosophical assumptions underlying scientific research, as well as research on educational policies for the teaching of school science. The purpose is to develop science curricula more in terms of a science of ourselves. Together, these trends define the need for a unifying core program in the sciences that will replace the traditional discipline approach. The professional education of teachers will also need to be modernized to meet the challenges of a public understanding of science in the context of a human science.

Social Inquiry

One of the oldest goals of science teaching has been the attempt to educate all students in how research scientists pursue new theories and advance progress in their field of interest (Lakatos & Musgrave, 1970). Seldom recognized is that there are many forms of inquiry in the sciences (Latour, 1987). Over the past quarter of a century, scientific inquiry as a primary purpose for teaching school science has been challenged—increasingly so by revolutionary changes in the practice of contemporary science.

The new face of science characterized by strategic research is focused on the utilization of science knowledge for the benefit of human welfare and social and economic progress. This shift in the primary purpose of scientific research and at the same time as the nation's move to a science/technology–oriented democracy has made social inquiry a purpose for education in the sciences at all educational levels (Barber, 1990; NSF Advisory Committee for Science Education, 1970).

Other events fostering social inquiry as a purpose for education in the sciences are (a) the nation's shift to a knowledge-intensive society and (b) the development of new concepts in the cognitive sciences focused on knowing, understanding, and utilization of science knowledge. There is a consensus that the sciences of today represent a break with the past and so must also be the case for public education in the sciences.

A sketch comparing social inquiry with scientific inquiry is the subject of the following paragraphs. Scientific inquiry represents a presumed procedure for the discovery of new knowledge or verification of existing knowledge about the natural world. It is focused upon the quantification of information. Social inquiry seeks to provide ways for closing the gap between the laboratory and the lived-in world of students. It depends on the findings of scientific inquiry but interprets the information in qualitative terms.

In social inquiry, whatever procedure is used, the result is qualitative in nature and likely subjective. Two students using the same database may get different answers: witness the ongoing interpretations in the U. S. Congress on health, welfare, and the environment. Social and scientific inquiry are alike in that both represent a cognitive framework for dealing with knowledge. However, the results from the two approaches are not the same. Social inquiry leads to policies about the utilization of scientific knowledge, while scientific inquiry leads to new theories about the natural world. The emergence of strategic research in the sciences essentially blends both approaches.

Social inquiry is not purely objective or fully predictable. Instead, its findings are subject to politics, economics, laws, morals, values, risks, and personal development. Race, gender, and cultural factors currently influence both social and science inquiry. It is within these contexts that citizens are required to make decisions or form judgments and take actions. Social inquiry thus becomes a process of logical or rational reasoning in terms of making science operational in human affairs. The process rests upon recognizing the reliability and validity of the knowledge to be used. In nearly all aspects of human endeavors there is a mass of superstition, myths, misconceptions, and bias to deal with. A common example of these factors is that of health as influenced by advertising, religion, and traditions.

Social inquiry serves to generate the big picture of a problem that can be analyzed for further critical investigations. The findings tend to be correlative, a blend of science and the humanities, culture, technology, and society. Social inquiry is a more difficult learning context than scientific inquiry, demanding what has come to be known as "higher order thinking skills," skills essential for the utilization of science knowledge in human affairs (Boulding & Senesh, 1983; Brim, 1969; Lynn, Jr., 1978; Moore, 1993; Wiesner, 1962). The skills are also those deemed essential in the world of work (Hurd, 1989b, pp. 15–40). Much of social inquiry like that of strategic research in the sciences is more fruitful if carried out by teams of individuals. Team investigations are a form of shared inquiry and serve to broaden each member's insights on a problem. An outcome of social inquiry is social literacy.

Adaptive Technology

A distinguishing characteristic of human evolution is the impact of technology on the adaptability and survival rate of our species. It is estimated that there are currently more than 500 technological devices that function in some way to extend the capacities of the human body to function as nature intended.

Canes, staffs, and walking sticks as adaptive aids are thousands of years old. Today, rehabilitation engineering has developed artificial limbs, some electronically controlled, that assure near normal walking capacities; new joints are also available. Various technologies such as eyeglasses and hearing aids improve our senses. One award-winning invention for 1996 was an ear implant designed to restore hearing after a patient's years of total deafness (Discover Awards, 1996). There are computers which extend human memories and "intelligent" robots that can challenge human thinking on the interpretation of data. Pacemakers and kidney machines improve organ function and also human survival rates. Gene therapies are now being tested to correct a variety of maladaptive inherited conditions of the body such as diabetes, the immune system, Huntington's chorea, certain cancers, and HIV ("Gene Therapies," 1995).

Research in gene therapy is not only a new field of research in biology but also a social and legal issue. There are human biologists and members of the general public who question whether scientists have the right to change the hereditary pattern of a human being and to what extent. What are the risks and benefits in changing the adaptive characteristics of a human being, a change that is permanent?

The field of adaptive technology is generally recognized as bioengineering. Among the developments are robotic arms and legs and artificial hips (Kemp, 1988, pp. 46–70). The field is shaped by researchers in physiology, pharmacology, mechanical and electrical engineering, molecular biology, neurochemistry, materials science, biophysics, and mathematics. Bioengineering is an excellent example of what is meant by "biology in a new vein," introduced in Chapter 6.

CHAPTER 9

Epilogue

This book emphasizes throughout the nature and ethos of science, the character of our society and economy, and how people live, work, and communicate. The biosciences now dominate the physical sciences as the center of research. Science has become a basis for social action in our culture and is becoming more a civic science. Strategic research is more socially driven than theory driven. Developments in contemporary science and technology are major elements in the nation's shift to a knowledge-based global economy. The sum of these changes has outmoded the rationale and goals underlying science curricula in schools and most colleges (AAAS, 1995; Griffiths, 1993; NSF, 1995).

The interaction of all these cultural shifts calls for a unifying perspective of science education that is oriented toward helping young people cope with life in the twenty-first century and that is supportive of a science/technology–based democracy. The current role of science in our culture demands cultural literacy and social inquiry as purposes for teaching science, fosters the building of human capital, and enables social and economic progress.

A number of scholars have commented on the need to link the natural and social sciences to evolve new perspectives of science education. Agnes E. Meyer, a philosopher, states her view as follows:

> We need a prophet of a brave new world, not like Huxley's but one that is really brave and really new. But before these geniuses can appear upon the scene the experts in the natural and social sciences, together with the humanists, must lay the groundwork by the same cooperative endeavor that animated the various scientific experts who split the atom. In short, we must achieve a humanism that is truly scientific and a science that is truly humane. (1957, pp. 4–5)

C. P. Snow, author of *The Two Cultures* (1959), at a meeting of the U. S. Committee on Science and Astronautics (January 26–27, 1966), described the problem as follows:

We need to know, more exactly, how we are living here and now. We are ignorant of the social life around us; we are more ignorant than is wise or safe or human. And this is where I come back to a plea for the mixed-up-ness of scientists, politicians, administrators, all the others—doctors, priests, citizens of goodwill—who are not cut off from our common humanity.

Years ago Margaret Mead, an anthropologist and a president of the American Association for the Advancement of Science, pointed out a crucial dimension to consider in planning institutional changes. She noted in a lecture at Harvard Business School that "No one will live all his life in the world into which he was born and no one will die in the world in which he worked in his maturity" (Mead, 1958, p. 34). In science teaching this means that students must know more than they are taught in their years of schooling, thus establishing lifelong learning as a goal of science teaching.

During the 1970s it became evident that the natural sciences and most fields of the social sciences needed to be reinterpreted and synthesized in order to reflect America's changing culture. The shifts in various aspects of our culture identify new purposes for education in general and especially in the natural and social sciences (Daedalus, 1977a, b, Vols. 1, 2).

In 1983 a special government committee, the National Commission on Excellence in Education, was appointed to examine various dimensions of reforming education in terms of fostering our democratic culture and the national life of citizens. Its report, *A Nation at Risk* (Gardner, 1983), notes that all efforts to bring science education into harmony with contemporary fields of scientific research, the social and economic progress of the nation, and the welfare of individuals have stalled. Scholars see the issue as a lack of any national agency to study the dimensions of such complex problems.

A major justification for the establishing of a National Academy of Education is that the production and utilization of knowledge has become the nation's and every individual's chief economic resource in the twenty-first century. Of the total mass of available knowledge, science/technology ranks at the top as an economic resource.

A NATIONAL ACADEMY OF EDUCATION IN THE SCIENCES

Scholars who have studied the reform of education in the sciences have noted the complexity of inventing curricula to meet the goals of developing human capital and assuring economic progress, set in the framework of a rapidly changing knowledge-intensive culture. One of the difficulties is that scientists and technologists, business representatives, citizens, education policy makers, science teachers, politicians, and cognitive researchers do not communicate or even speak the same language. Consequently, little progress has been made over the past 50 years in creating a core science curriculum required of all students which recognizes cultural shifts and the foreseeable demands of life in the twenty-first century. This task is beyond the capacity of isolated research committees seeking to resolve one educational problem at a time. Scholars who have studied the situation see a need for a permanent national agency with responsibility for exploring all aspects of educational reform in a unifying context.

The Committee on Education and Labor, House of Representatives, has proposed the creation of a National Institute of Education. An important task of this institute would be "the continual redefinition of purpose, and this is a field in which there can be significant disagreement" (H. R. 3606, Appendix 2, 1972, p. v). In its appendix 3, the Hearings on H. R. 3606 focus on "alternative futures in American education." The central perspective for this volume becomes an issue for the National Institute of Education to consider in terms of our changing society, economic patterns, and culture of science. Dealing adequately with reform issues would require a diverse combination of scholars and specialists working together over a span of years (H. R. 3606, Appendix 3, 1972, Hurd, 1971b).

An extensive study by the U.S. National Research Center on Science and Mathematics Education in the United States and a number of foreign countries found there is a "splintered vision" of what a modern education in the sciences should mean. It was observed that the United States is more than a quarter of a century into the science education reform movement with little progress having been made. In their report *A Splintered Vision: An Investigation of U.S. Science and Mathematics Education* (Schmidt, McKnight, & Raizen, 1997), the committee concluded:

> There is no one at the helm of U.S. mathematics and science education. In truth, there is no one helm. No single coherent vision of how to educate today's children dominates U.S. educational practice in either science or mathematics. There is no single, commonly accepted place to turn to for such visions. The visions that shape U.S. mathematics and science education are splintered. This is seen in what is planned to be taught, what is in textbooks, and what teachers teach. (p. 1)

In 1985 the National Research Council (NRC) noted the need for a reform of science education in terms of a larger mission "than currently existed" (NRC, 1985, p. v). NRC supported the position of the National Science Board Commission on Precollege Education in Mathematics, Science, and Technology:

> ... improved preparation of all students in the fields of mathematics, science, and technology is essential to the maintenance and development of our nation's

economic strength, to its military security, to its commitment to the democratic ideal of an informed and participating citizenry, and to fulfilling personal lives for its people. (National Science Foundation, 1983, p. 1)

In 1988 the NRC conducted a second study of K–12 science education in the United States. The purpose of each study was to develop indicators of the present "quality of science education" in the United States. The lack of national goals and a national curriculum framework creates a demand for indicators to assess just what is being accomplished in the teaching of school science. In both reports the final statements on indicators were the opinions of an individual or a series of small committees (NRC, 1988). The findings in the two books are more symptoms of problems in science education rather than a national plan for their resolution.

FRAMING A NATIONAL ACADEMY OF SCIENCE EDUCATION

Since the 1960s the pressures for a reformation of science education have consistently yielded few results beyond that of confusion. There has been a lack of cohesiveness in both national and local efforts. The reform movement has been propelled for the most part by a wave of criticisms of schools, teachers, and curricula. On one hand there are pressures to "get back to basics." On the other, the demand is for an education in science that responds to the growing impact of science and technology on the course of our democratic society and the quality of life.

While there is much written on each of these topics, there is a lack of serious study to isolate the broader meaning of each of these attributes and their interactions with other factors. Without the benefit of intellectual nourishment and relevant research, each attribute ends up as a slogan. Most of the current slogans have been around for hundreds, some even thousands, of years. It was 3000 years ago that Confucius recognized "the value of knowledge is in its use." Today, the movement in science education is "the utilization of knowledge." Although a slogan may be recognized, curriculum standards are not written in terms likely to ensure that what is being taught in science/technology meets social demands and adds to human capital. Note again the comments of Oppenheimer and Einstein in the prologue of this book. The reform movement requires new ways of thinking about curricula, instruction, and learning (National Association for Research in Science Teaching, 1996).

To deal with the mismatch between science education and our culture requires the concerted intellectual resources of a wide variety of knowledgeable scholars. It also requires diversified interdisciplinary research teams. A model for this approach is found in the conduct of strategic research in the natural sciences. Typically, a study team would consist of philosophers, sociologists of science, cognitive scientists, economists, teachers, science educators, educators who specialize in policy, and representatives from the U.S. Congress, National Academy of Sciences, National Science Foundation, and American Association for the Advancement of Science. Also included would be private foundations that focus on youth development and the economics of science/technology, representatives from the world of work, scientific societies, and other specialists, depending on the problem to be studied.

A major concern in the reform movement is the competitive position of the United States in a global knowledge-intensive economy influenced by achievements in science and technology. One arm of an academy should focus on science education reforms in foreign countries. Most of the developed countries worldwide are now in the process of rethinking their education in the sciences (UNESCO-NIER, 1996, pp. 1, 3; Black & Atkin, 1996).

Whatever task an academy team is assigned, it must accomplish more than simply to identify a problem or issue, and include a conceptual consensus of its status and needed research. At some point, grades K–12 science education issues should be examined in terms of implied responsibilities of colleges and universities for the long-term resolution of the issues (NSF, 1996, S. Rep. No. 3095, 1990; S. Rep. No. s.2., 1992).

The academy will need a permanent staff of 15 to 30 specialists and a rotating staff that deals with the details of special problems. The first responsibility of the permanent staff is to plan the organization of the academy and develop some notion of where we are in the reform process. There are hundreds of national reports and congressional actions that need to be analyzed, synthesized, and interpreted. It is essential that the findings of any team or committee be made available to the teaching community and the public. Most of all, they should be written in a language that science teachers and parents can easily understand.

The nearest model we have for the tasks of an academy of science education is the Morrill Act (1862), leading to the establishment of land grant colleges. The general plan was to conduct agricultural research and deliver the findings of research to farmers by means of county agents. There is little question but that this was the single most productive economic investment in the history of the United States.

Today and for the forseeable future, the nation again faces an economic crisis, one rooted in the productive knowledge of science/ technology. We now lack a national framework in science education that offers a reasonable program that could improve the nation's economic achievements as well as the quality of life. This list of responsibilities should not be viewed as comprehensive, but simply a thread of ideas to initiate early actions of the academy.

References

Abelson, P. H. (1986). Instrumentation and computers. *American Scientist, 2,* 182–192.

Adolescence in the 1990s: Risk and opportunity. (1993). *Teachers College Record, 94*(3).

Aikenhead, G. S. (1985). Collective decision making in the social context of science. *Science Education, 69*(4), 453–475.

Alexander, L. (1991). *America 2000: An education strategy.* Washington, DC: U.S. Department of Education.

American Association for the Advancement of Science (AAAS). (1957). Committee report on the social aspects of science. *Science, 125,* 13–47.

American Association for the Advancement of Science (AAAS). (1990). *Science for all Americans.* New York: Oxford University Press.

American Association for the Advancement of Science (AAAS). (1992). The third branch of science. *Science, 256,* (62).

American Association for the Advancement of Science (AAAS). (1993). *Benchmarks for science literacy.* New York: Oxford University Press.

American Association for the Advancement of Science (AAAS). (1995). *1995 annual report.* Washington, DC: Author.

American Medical Association (AMA). (1986). *AMA white paper on adolescent health.* Washington, DC: Author.

American Psychological Association (APA). (1993). *Violence and youth: Psychology's response.* Washington, DC: Author.

Amour, R. (1963). *Through darkest adolescence.* New York: McGraw-Hill.

Ashby, W. R. (1963). *Introduction to cybernetics.* New York: John Wiley and Sons.

ASTC Newsletter. (1995, March/April). Survey of Science Education Centers.

Atkinson, R. C., & Jackson, G. B. (Eds.). (1977). *Research and education reform: Roles for the Office of Education Research and Improvement.* Washington, DC: National Academy Press.

August, D. (1989). *Survey of state policies and progress for the middle grades.* Washington, DC: Children's Defense Fund.

Ayala, F. (1989). *On being a scientist.* Washington, DC: National Academy Press.

Bandura, A. (1977). *Social learning theory.* Englewood Cliffs, NJ: Prentice-Hall.

Barber, B. (1990). *Social studies of science.* New Brunswick, NJ: Transaction.

Barnard, J. D. (Ed.). (1960). *Rethinking science education, the 59th yearbook of the National Society for the Study of Education.* Part 1. Chicago: University of Chicago Press.

Barton, P. E., & Cooley, R. J. (1992). *America's smallest school: The family*. Princeton, NJ: Educational Testing Service Policy Center.

Bauer, H. H. (1992). *Scientific literacy and the myth of the scientific method*. Urbana, IL: University of Illinois Press.

Beane, J. A. (1990). *A middle school curriculum: From rhetoric to reality*. Columbus, OH: National Middle School Association.

Beane, J. A., & Lipka, R. P. (1986). *Self-concept, self-esteem, and the curriculum*. New York: Teachers College Press.

Beauchamp, W. L. (1932). *Instruction in science* [Bulletin No. 17; National Survey of Secondary Education, Monograph No. 22]. Washington, DC: U.S. Superintendent of Documents.

Becker, E. (1971). *The lost science of man*. New York: George Braziller.

Berkner, L. V. (1964). *The scientific age: The impact of science on society*. New Haven, CT: Yale University Press.

Berkovitz, I. H., & Sugar, M. (1986). Indications and contradictions for adolescent group psychotherapy. In M. Sugar (Ed.), *The adolescent in group and family therapy*. Chicago: University of Chicago Press.

Bernal, J. D. (1939). *The social function of science*. New York: Macmillan.

Black, P., & Atkin, J. M. (1996). *Changing the subject: Innovations in science, mathematics and technology education*. New York: Routledge.

Boulding, K. E. (1964). *The meaning of the twentieth century: The great transition*. New York: Harper and Row.

Boulding, K. E., & Senesh, L. (1983). *The optimum utilization of knowledge: Making knowledge serve human betterment*. Boulder, CO: Westview Press.

Branscomb, L. (1981). *Only one science: Twelfth annual report of the National Science Board*. Washington, DC: U.S. Superintendent of Documents.

Brauman, J. I. (1993). Instrumentation. *Science, 260*, 1407.

Brim, O. G., Jr. (1969). *Knowledge into action: Improving the nation's use of the social sciences*. Washington, DC: National Science Foundation.

Brzezinski, Z. (1970). *Between two ages: America's role in the technetronic era*. New York: Viking Press.

BSCS (1995). *Designing the science curriculum: A report on the implications of standards for science education*. Colorado Springs, CO: Author.

Buck, P. H. (1945). *Report of the Harvard Committee: General education in a free society*. Cambridge, MA: Harvard University Press.

Burd, S. (1994). U. W. won't back creation of human embryos for research. *Chronicle of Higher Education, 41* (16), A32.

Bybee, R. W. (1996). *National standards and the science curriculum: Challenges, opportunities, and recommendations*. Dubuque, IA: Kendall/Hunt.

Caldwell, O. W. (1920). *Reorganization of science in secondary schools*. Bulletin No. 26. Bureau of Education. Washington, DC: U.S. Government Printing Office.

California State Department of Education. (1987). *Caught in the middle: Educational reform for young adolescents in California schools*. Sacramento, CA: Author.

Campbell, C. A., Jr., & Romer, R. (1992). *Raising standards for American education*. Washington, DC: U.S. Government Printing Office.

Carnegie Commission on Science, Technology, and Government. (1988). *Science and technology and the President*. New York: Author.

Carnegie Commission on Science, Technology, and Government. (1991). *In the national interest: The federal government in the reform of K–12 math and science education*. New York: Author.

Carnegie Commission on Science, Technology, and Government. (1992a). *Science, technology, and the states in America's third century*. New York: Author.

Carnegie Commission on Science, Technology, and Government. (1992b). *Enabling the future: Linking science and technology to societal goals*. New York: Author.

Carnegie Commission on Science, Technology, and Government. (1992c). *A science and technology agenda for the nation: Recommendations for the President and Congress*. New York: Author.

Carnegie Commission on Science, Technology, and Government. (1993a). *Science, technology, and government for a changing world*. New York: Author.

Carnegie Commission on Science, Technology, and Government. (1993b). *Science and technology in judicial decision making: Creating opportunities and meeting challenges*. New York: Author.

Carnegie Council on Adolescent Development. (1989). *Turning points: Preparing youth for the twenty-first century*. New York: Author.

Carnegie Council on Adolescent Development. (1995). *Great transitions: Preparing adolescents for a new century*. Waldorf, MD: Author.

Center for Early Adolescence. (1988). *Before it's too late: Dropout prevention in the middle grades*. Carrboro, NC: Author.

Center for Educational Research and Evaluation. (1981). *The status of middle school and junior high school science*. Colorado Springs, CO: Biological Sciences Curriculum Study.

Center for the Future of Children. (1995a). *The future of children: Critical issues for children and youths*. Los Altos, CA: David and Lucile Packard Foundation.

Center for the Future of Children. (1995b). *The future of children: Long-term outcomes of early childhood programs*. Los Altos, CA: David and Lucile Packard Foundation.

Chalk, R. A. (1988). *Science, technology, and society: Emerging relationships*. Washington, DC: American Association for the Advancement of Science.

Children's Defense Fund. (1988). *Making the middle grades work*. Washington, DC: Author.

CIBA Foundation Symposium. (1972). *Civilization and science: In conflict or collaboration?* New York: Associated Scientific.

Clinton, B., & Gore, A. (1992). *Putting people first: How we can all change America*. New York: Times Books.

Clinton, W. J., & Gore, A. (1993). *Technology for America's economic growth: A new direction to build economic strength*. Washington, DC: U.S. Government Printing Office.

Clinton, W. J., & Gore, A. (1994). *Science in the national interest*. Washington, DC: U.S. Government Printing Office.

Cohen, J. (1994). Frontiers in medicine. *Science, 265*.

Cohen, Y. A. (1968a). *Man in adaptation: The cultural present.* Chicago: Aldine.

Cohen, Y. A. (1968b). *Man in adaptation: The biosocial background.* Chicago: Aldine.

Collins, W. A. (1984). *Development during middle childhood: The years from six to twelve.* Washington, DC: National Academy Press.

Committee of Associated Institutions of Science and Arts. (1861). *Objectives and plan for an institute of technology.* Boston: John Wilson & Son. (Photocopy courtesy of Institute Archives, MIT)

Conference Report to accompany H. R. 1804, House of Representative, March 21, 1994. Washington, DC: U.S. Government Printing Office.

Corney, R. (1972). *The human agenda.* New York: Simon and Schuster.

Daedalus. (1971, Fall). *Twelve to sixteen: Early adolescence.* Cambridge, MA: American Academy of Arts and Sciences.

Daedalus. (1977a, Summer, special issue). *Discoveries and interpretations studies in contemporary scholarship.* Vol. 1. Cambridge, MA: Academy of Arts and Sciences.

Daedalus. (1977b, Fall). *Discoverers and interpretations studies in contemporary scholarship.* Vol. 2. Cambridge, MA: American Academy of Arts and Sciences.

David and Lucile Packard Foundation, Center for the Future of Children. (1995, Summer/Fall). *The Future of Children: Critical issues for children and youths,* 5(2). Los Altos, CA: Author.

de Nemours, D. (1923). *National education in the United States.* Newark, DE: University of Delaware.

Department of Education. (1996). *Achieving the goals: Goal 1, all children in America will start school ready to learn.* Washington, DC: Superintendent of Documents.

Dick, H. (Ed.). (1955). *Selected writings of Francis Bacon.* New York: Random House.

Discover Awards. (1996, July). *Discover. New York Times Supplement.* New York: New York Times.

Doby, J. T., Boskoff, A., & Pendleton, W. W. (1973). *Sociology: The study of man in adaptation.* Lexington, MA: D.C. Heath.

Doyle, D. A., & Hartle, T. W. (1985). *Excellence in education: The states take charge.* Washington, DC: American Enterprise Institute for Public Policy Research.

Drucker, P. F. (1968). *The age of discontinuity: Guidelines to our changing society.* New York: Harper and Row.

Druckman, D., & Bjork, R. A. (1994). *Learning, remembering, believing: Enhancing human performance.* Washington, DC: National Academy Press.

DuBridge, L. A. (1959). *Education for the age of science.* Washington, DC: The White House.

Duschl, R. A. (1985). Science education and philosophy of science: Twenty-five years of mutually exclusive development. *School Science and Mathematics, 85,* 541–555.

Duschl, R. A. (1986). The changing concept of scientific observation. In *1985 NSTA Yearbook: Science, Technology, Society.* Washington, DC: National Science Teachers Association.

Duschl, R. A. (1988). Abandoning the scientist's legacy of science education. *Science Education, 72,* 55–62.

Eddy, E. D. (1956). *Colleges for our land and time: The land-grant idea in American education.* New York: Harper & Brothers.

Engelhardt, H., Jr., & Callahan, D. (Eds.). (1978). *Morals, science, and sociality.* Hastings-on-Hudson, NY: Institute of Society, Ethics, and the Life Sciences.

England, J. M. (1982). *A patron for pure science: The National Science Foundation's formative years, 1945–1957.* Washington, DC: National Science Foundation.

Environmental Studies Division, Office of Research and Monitoring. (1973). *The quality of life concept: A potential new tool for decision-makers.* Washington, DC: U.S. Environmental Protection Agency.

Federal Coordination Council for Science, Engineering, and Technology Committee on Education and Human Resources. (1991). *By the year 2000: First in the world.* Vols. 1, 2. Washington, DC: Author.

Feldman, S. S., & Elliott, G. R. (1990). *At the threshold: The developing adolescent.* Cambridge, MA: Harvard University Press.

Fletcher, J. C., Miller, F. G., & Caplan, A. L. (1994). Facing up to bioethical decisions. *Issues in Science and Technology, 11.*

Ford, W. D. (1991). *Education 2005: The role of research and development in an overwhelming campaign for education in America.* Washington, DC: U.S. Government Printing Office.

Frankel, M. S., & Teich, A. H. (1974). *The genetic frontier: Ethics, law and policy.* Washington, DC: American Association for the Advancement of Science.

Franklin, B. (1743). *A proposal for promoting useful knowledge among the British plantations in America.* (Photocopy of the original manuscript, courtesy of Yale University Library)

Fuchs, V. R., & Reklis, D. M. (1992). America's children: Economic perspectives and policy options. *Science, 255* (5040), 41.

Furby, L., & Beyth-Maron, R. (1990). *Risk taking in adolescence.* Washington, DC: Carnegie Council on Adolescent Development.

Gabor, D. (1964). *Inventing the future.* New York: Alfred A. Knopf.

Gardner, D. P. (1983). *A nation at risk: The imperative for educational reform.* Washington, DC: U.S. Government Printing Office.

Gene Therapies. Growing pains. (1995). Special News Report in Science. *Science, 69* (5527), 1050–1053. Washington, DC: American Association for the Advancement of Science.

George, M. D. (1996). *Shaping the future: New expectations for undergraduate education in science, mathematics, engineering, and technology.* Arlington, VA: National Science Foundation.

Gibson, K. R., & Ingold, T. (1994). *Tools, language, and cognition in human evolution.* New York: Cambridge University Press.

Glass, B. (1970). *The timely and the timeless: The interrelationships of science, education, and society.* New York: Basic Books.

Glatthorn, A. A. (1987). *Curriculum leadership.* Glenview, IL: Scott, Foresman.

Gleick, J. (1988). *Chaos: Making a new science.* New York: Penguin Books.

Goals 2000: Educate America Act. (1994). Public Law 103-227. 100d Congress. Washington, DC: U.S. Government Printing Office.

Gore, A. (1996). The metaphor of distributed intelligence. *Science, 272* (5259), 177.

Grant, W. T. Foundation (1988). *The forgotten half: Pathways to success for American youth and young families.* Washington, DC: Author.

Graubard, S. R. (Ed.). (1973). The search for knowledge. *Daedalus, 102*, 2. Cambridge, MA: American Academy of Arts and Sciences.

Greenwood, A., Bartusiak, M. F., Burke, B. A., & Edelson, E. (1992). *Science at the frontier: Vol. 1*. Washington, DC: National Academy Press.

Griffiths, P. A. (1993). *Science, technology, and the federal government: National goals for a new era*. Washington, DC: National Academy Press.

Grinnell, F. (1992). *The scientific attitude* (2d ed.). New York: Guilford Press.

Gross, P. R., & Levitt, N. (1994). *Higher superstition: The academic and its quarrels with science*. Baltimore: Johns Hopkins University Press.

H. R. 1804. (1993). *Goals 2000: Educate America Act*. 103d Cong., 1st Sess. (1993). Washington, DC: U.S. Government Printing Office.

H. R. 2884. *Conference report. School-to-work opportunities act* (1994). 103d Cong. To accompany H. R. 2884. Report No. 239, May 4, 1994. Washington, DC: U.S. Government Printing Office.

H. R. 3606. Appendix 2. (1972). *Purpose and process. Readings in educational research and development*. Washington, DC: U.S. Government Printing Office.

H. R. 3606. Appendix 3. (1972). *Alternative futures in American education*. Washington, DC: U.S. Government Printing Office.

H. R. 4323. 102d Cong. (A Bill). 2d Sess. (1992).

Hall, G. S. (1905). *Adolescence: Its psychology and its relations to physiology, anthropology, sex, crime, religion, and education*. Vols. 1, 2. New York: Appleton.

Hall, S. S. (1992). How technique is changing science. *Science, 257*, 344–349.

Hamburg, B. A. (1990). *Life skills training: Preventive interventions for young adolescents*. Washington, DC: Carnegie Council on Adolescent Development.

Hamburg, D. A. (1987). *Preparing for life: The critical transition*. New York: Carnegie Corporation.

Hamburg, D. A. (1993). *Preventing contemporary intergroup violence*. New York: Carnegie Corporation.

Hamburg, D. A. (1995). *Great transitions: Preparing adolescents for a new century. Concluding report*. New York: Carnegie Corporation.

Hanushek, E. A. (1994). *Making schools work: Improving performance and controlling costs*. Washington, DC: The Brookings Institute.

Hanushek, E. A., & Jorgenson, D. W. (Eds.). (1996). *Improving America's schools: The role of incentives*. Washington, DC: National Academy Press.

Hardin, G. (1959). *Nature and man's fate*. New York: Rinehart.

Harrison, G. A. (1993). *Human adaptation*. New York: Oxford University Press.

Hechinger, F. M. (1992). *Fateful choices: Healthy adolescents for the twenty-first century*. New York: Hill and Wang.

Hendee, W. R. (Ed.). (1991). *The health of adolescents: Understanding and facilitating biological, behavioral, and social development*. San Francisco: Jossey-Bass.

Hoke, F. (1993). Medical informatics: Where life sciences, computer science converge. *The Scientist, 7*, 1, 7.

Hoke, F. (1994a). Science in the courtroom: What evidence is admissible—and who decides? *The Scientist, 8*(1), 4–5.

Hoke, F. (1994b). Scientists and lawyers: Projects aim to bridge gap between the traditional contentious professions. *The Scientist, 8*(1), 4–5.

Holland, L. (1990). Which scientists might be honored with the Nobel prize? *The Scientist, 4* (19), 18.

Holton, G. (1975). Scientific optimism and social concerns. *Hastings Center Report 5* (6), 39–47.

Holton, G. (1993). *Science and anti-science.* Cambridge, MA: Harvard University Press.

Hurd, P. D. (1958). Scientific literacy: Its meaning for American schools. *Educational Leadership, 16* (1), 13–16.

Hurd, P. D. (1969a). Guidelines for development of a life science program in the middle school. *BSCS Newsletter,* No. 34, 2–8.

Hurd, P. D. (1969b). *New directions in teaching secondary school science.* Chicago: Rand McNally.

Hurd, P. D. (1970a). *New curriculum perspectives for junior high school science.* Belmont, CA: Wadsworth.

Hurd, P. D. (1970b). Scientific enlightenment for an age of science. *The Science Teacher, 37,* 13–15.

Hurd, P. D. (1971a). Biology as a study of man and society. *The American Biology Teacher, 33* (10), 399–400, 408.

Hurd, P. D. (1971b). Research in science education: Planning for the future. *Journal of Research in Science Teaching, 8* (3), 243–249.

Hurd, P. D. (1978a). The BSCS human sciences project. *BioScience, 28* (1), 36–38.

Hurd, P. (1978b). *Early adolescence perspectives and recommendations.* Washington, DC: National Science Foundation.

Hurd, P. D. (1983). State of precollege education in mathematics and science. *Science Education, 67*(1), O57–O67.

Hurd, P. D. (1984). *Reforming science education: The search for a new vision.* Washington, DC: Council for Basic Education.

Hurd, P. D. (1985). Science education for a new age. *The National Association of Secondary School Principals (NASSP) Bulletin, 69,* 83–92.

Hurd, P. D. (1986). Perspectives for the reform of science education. *Phi Delta Kappan, 67,* 353–358.

Hurd, P. D. (1987). A nation reflects: The modernization of science education. *Bulletin of Science, Technology, & Society, 7,* 9–13.

Hurd, P. D. (1989a). A life science core for early adolescents. *Middle School Journal, 20* (5), 20–23.

Hurd, P. D. (1989b). Science education and the nation's economy. In A. B. Champagne, B. E. Lovitts, & B. J. Calinger (Eds.), *This year in school science: Scientific literacy.* Washington, DC: American Association for the Advancement of Science.

Hurd, P. D. (1990). Change and challenge in science education. *Journal of Research in Science Teaching, 27,* 413–414.

Hurd, P. D. (1991a). Closing the educational gaps between science, technology, and society. *Theory Into Practice, 30,* 251–259.

Hurd, P. D. (1991b). Why we must transform science education. *Education Leadership, 49*, 33–35.

Hurd, P. D. (1993). Comment on science education research: A crisis of confidence. *Journal of Research in Science Teaching, 30* (8), 1009–1011.

Hurd, P. D. (1994a). Technology and the advancement of knowledge in the sciences. *Bulletin of Science, Technology, & Society, 14*, 127–135.

Hurd, P. D. (1994b). New minds for a new age: Prologue to modernizing the science curriculum. *Science Education, 78*, 103–116.

Hurd, P. D. (1995). Reinventing the science curriculum: Historical reflections and new directions. In *Redesigning the science curriculum: A report on the implications of standards and benchmarks for science education* (R. M. Bybee and J. D. McInerney, Eds.). Colorado Springs, CO: Biological Science Curriculum Study.

Hurd, P. D., & Gallagher, J. J. (1968). *New directions in elementary science teaching*. Belmont, CA: Wadsworth.

Hutchins, R. M. (1968). *The learning society*. New York: Praeger.

Institute for Scientific Information (ISI). (1993). Scientific papers: Top producers of 1991. *Science, 159*, 180.

Jasanoff, S., Markle, G. E., Peterson, J. C., & Pinch, T. (Eds.). (1994). *Handbook of science and technology studies*. Thousand Oaks, CA: Sage.

Jennings, J. F. (1995). *National issues in education: Goals 2000 and school to work*. Bloomington, IN: Phi Delta Kappa International.

Jewett, B., & Butcher, J. H. (1964). *Aristotle: politics and poetics. Book 8*. New York: Heritage Press.

Johnson, D. W., & Johnson, R. T. (1995). *Reducing school violence through conflict resolution*. Alexandria, VA: Association for Supervision and Curriculum Development.

Kahn, P. (1994). Genetic diversity project tries again. *Science, 266*, 720–722.

Katzenmeyer, C. (1979). Focus on early adolescence: A new emphasis for the Science Education Directorate. *Science Education, 63* (1), 139–142.

Kelly, J. K., Carlsen, W. S., & Cunningham, C. M. (1993). Science education in sociocultural context: Perspectives from the sociology of science. *Science Education, 77*, 207–220.

Kemp, M. (1988, November). Rebuilding the body. *Discover*.

Kerr, R. A. (1991). Chemistry with a thousand faces. *Science, 253*, 1212.

Lakatos, I., & Musgrave, A. (1970). *Criticism and the growth of knowledge*. New York: Cambridge University Press.

Lane, N. (1996a). Wanted: Citizen-scientists. *Science, 271* (5252), 904.

Lane, N. (1996b). Science and the American dream. *Science, 271* (5252), 1037.

Lanham, U. (1968). *Origins of modern biology*. New York: Columbia University Press.

Latour, B. (1987). *Science in action: How to follow scientists and engineers through society*. Cambridge, MA: Harvard University Press.

Layton, D. (1973). *Science for the people*. New York: Neal Watson Academic Publications.

Lenoir, T. (1988). Practice, reason, contexts: The dialogue between theory and practice. *Science in Context, 2*, 5–22.

Lewenstein, B. (Ed.). (1992). *When science meets the public.* Washington, DC: American Association for the Advancement of Science.

Lewis, A. C. (1993). *Changing the odds: Middle school reform in progress, 1991–1993.* New York: Edna McConnell Clark Foundation.

Lipsitz, J. (1977). *Growing up forgotten.* Lexington, MA: Lexington Books.

Lipsitz, J. (1983). *Successful schools for young adolescents.* New Brunswick, ME: Transaction Books.

Longino, H. E. (1990). *Science as social knowledge.* Princeton, NJ: Princeton University Press.

Lounsbury, J. H. (1991). *As I see it.* Columbus, OH: National Middle School Association.

Lounsbury, J. H., & Clark, D. C. (1990). *Inside grade eight: From apathy to excitement.* Reston, VA: National Association of Secondary School Principals.

Lynn, L. E., Jr. (1978). *Knowledge and policy: The uncertain connection.* Washington, DC: National Academy of Sciences.

Maarshalk, J. (1992). Paradigm shifts in the quest for a science of science education. *The History and Philosophy of Science in Education: Vol. 2* (pp. 81–86). Kingston, ONT: Queen's University.

Machiavelli, N. (1977). *The prince* (J. B. Atkinson, Ed.). Indianapolis, IN: Bobbs-Merrill. p. 75. (Originally published in 1513)

Machlup, F. (1962). *The production and distribution of knowledge in the United States.* Princeton, NJ: Princeton University Press.

Manen, M. V. (1990). *Researching lived experience: Human science for an action sensitive pedagogy.* New York: State University of New York Press.

March, J. G. (Chair). (1985). *Mathematics, science, and technology education: A research agenda.* Washington, DC: National Academy Press.

Marshall, R., & Tucker, M. (1992). *Thinking for a living: Education and the wealth of nations.* New York: Basic Books.

Martin, B., Kass, H., & Wytze, G. (1990). Authentic science: A diversity of meanings. *Science Education, 74,* 541–554.

McElroy, W. D. (1970). *Science education: The task ahead for the National Science Foundation* (Report No. 71–13). Washington, DC: National Science Foundation.

Mead, M. (1958, November). *Harvard Business Review.*

Mead, M. (1976). Towards a human science. *Science, 191,* 903.

Metropolitan Life Survey. (1996). *Students voice their opinions on: Violence, social tension and equality among teens* (Part I). New York: Metropolitan Life Insurance Co.

Meyer, A. E. (1957). *Education for a new morality.* New York: Macmillan.

Michael, D. M. (1968). *The unprepared society: Planning for a precarious future.* New York: Basic Books.

Millstein, S. G., Peterson, A. C., & Nightingale, E. O. (Eds.). (1993). *Promoting the health of adolescents: New directions for the twenty-first century.* New York: Oxford University Press.

Moffat, A. S. (1993). New meetings tackle the knowledge conundrum. *Science, 259* (1253), 55.

Moore, J. A. (1993). *Science as a way of knowing: The foundations of modern biology.* Cambridge, MA: Harvard University Press.

Morell, V. (1996). Using science to help shape the nation's policies. *Science, 271* (5254), 1439.

Morganett, R. (1990). *Skills for living.* Champaign, IL: Research Press.

National Academy of Sciences (NAS). (1977). *Science: An American bicentennial view.* Washington, DC: Author.

National Academy of Sciences (NAS). (1986). *Research briefings, 1986.* Washington, DC: National Academy Press.

National Academy of Sciences (NAS). (1994). *National science education standards* (draft). Washington, DC: National Academy Press.

National Academy of Sciences (NAS). (1995). *Reshaping the graduate education of scientists and engineers.* Washington, DC: National Academy Press.

National Academy of Sciences, National Academy of Engineering, & Institute of Medicine. (1993). *Science, technology, and the federal government: National goals for a new era.* Washington, DC: National Academy Press.

National Association for Research in Science Teaching. (1996). Establishing a theoretical frame [Editorial]. *Journal of Research in Science Teaching, 33* (3), 225–227.

National Center for Education Statistics. (1996). *Youth indicators 1996: Trends in the well-being of American youth.* Washington, DC: U.S. Government Printing Office.

National Council on Education Standards and Testing. (1992). *Raising standards for American education: A report to Congress, the Secretary of Education Goals Panel, and the American people.* Washington, DC: Author.

National Education Association. (1894). *Report of the Committee of Ten on secondary school studies.* New York: American Book.

National Education Goals Panel. (1991). *National Education Goals Panel report: Building a nation of learners.* Washington, DC: Author.

National Education Goals Panel (1995). (Updated). *The national education goals report: Building a nation of learners.* Washington, DC: U.S. Government Printing Office.

National Issues Forums. (1994). *Kids who commit crimes.* Dubuque, IA: Kendall/ Hunt.

National Middle School Association (NMSA). (1995). *This we believe.* Columbus, OH: Author.

National Research Council (NRC). (1978). *Knowledge and policy: The uncertain connection.* Washington, DC: National Academy of Sciences Press.

National Research Council (NRC). (1988). *Improving indicators of the quality of science and mathematics education in grades K–12.* Washington, DC: National Academy Press.

National Research Council (NRC). (1993). *Losing generations: Adolescents in high-risk settings.* Washington, DC: National Academy Press.

National Research Council (NRC). (1996). *National science education standards.* Washington, DC: National Academy Press.

National Research Council Panel on School Science, Commission on Human Resources. (1979). *The state of school science.* Washington, DC: Author.

National Science Foundation (NSF). (1979). *Unanticipated benefits from basic research.* Washington, DC: Author.

National Science Foundation (NSF). (1980). *What are the needs in precollege science, mathematics, and social science education?* (SE–80–9). Washington, DC: Author.

National Science Foundation (NSF). (1982). *Emerging issues in science and technology.* Washington, DC: Author.

National Science Foundation (NSF). (1983a). *Educating Americans for the 21st century: A report to the American people and the National Science Board.* Washington, DC: Author.

National Science Foundation (NSF). (1983b). *Emerging issues in science and technology, 1981.* Washington, DC: Author.

National Science Foundation (NSF). (1991). *Science and engineering: Research benefits.* Washington, DC: Author.

National Science Foundation (NSF). (1995). *NSF in a changing world: The National Science Foundation's strategic plan.* Arlington, VA: Author.

National Science Foundation (NSF). (1996). *Shaping the future: New expectations for undergraduate education in science, mathematics, engineering, and technology.* Washington, DC: Author.

National Science Foundation (NSF) Advisory Committee for Science Education. (1970). *Science education: The task ahead for the National Science Foundation* (71–13). Washington, DC: Author.

National Science Teachers Association. (1992). *The content core: A guide for curriculum designers.* Vol. 1. Washington, DC: Author.

Nelkin, D., & Trancino, L. (1994). *Dangerous diagnostics: The social power of biological knowledge.* Chicago: The University of Chicago Press.

NSTA Committee on Scope, Sequence and Coordination. (1992). *The content core: A guide for curriculum designers.* Washington, DC: The National Science Teachers Association.

NW Education. (1996). *Mid kids: Learning in the middle years.* Portland, OR: Northwest Regional Education Laboratory.

Office of Educational Research and Improvement. (1991). *Youth indicators 1991: Trends in the well-being of American youth.* Washington, DC: U. S. Department of Education.

Office of Technology Assessment, Congress of the United States. (1995). *Learning to work: Making the transition from school to work.* Washington, DC: U.S. Government Printing Office.

Olson, S. (1986). *Biotechnology: An industry comes of age.* Washington, DC: National Academy Press.

Oppenheimer, J. R. (1947). *Physics in the contemporary world.* Cambridge, MA: MIT Press.

Oppenheimer, J. R. (1963). *Prospects in the arts and sciences* [A sound recording]. Boulder, CO: University of Colorado.

Organization for Economic Co-Operation and Development (OECD). (1996). *Lifelong learning for all.* Paris, France: Publications Service OECD.

Pabst, D. (1994). Changing the culture of science in research universities. *Science, 266* (5183), 669.

Panel on High-Risk Youth. (1993). *Losing generations: Adolescents in high risk settings.* Washington, DC: National Academy Press.

Panel on Understanding and Preventing Violence. (1993). *Understanding and preventing violence.* Washington, DC: National Research Council.

Penner, L. A., Batsche, G. M., Knoff, H. M., & Nelson, D. L. (Eds.). (1994). *The challenge in mathematics and science education: Psychology's response.* Washington, DC: American Psychological Association.

Powers, S. R. (1932). *A program for teaching science, the thirty-first yearbook.* Part 1. National Society for the Study of Education. Bloomington, IL: Public School.

Price, D. D. (1983). *Sealing wax and string.* George Sarton Memorial Lecture. AAAS Annual Meeting, Detroit. Glendale, CA: Mobiltape Co.

Project 2061: American Association for the Advancement of Science. (1990). *Science for all Americans.* New York: Oxford University Press.

Project 2061: American Association for the Advancement of Science. (1993). *Benchmarks for science literacy.* New York: Oxford University Press.

RANN 2. (1976). *Realizing knowledge as a resource.* (Vols. 1–6). Washington, DC: National Science Foundation.

Ray, J. D., & Nelson, G. E. (1971). *What a piece of work is man: Introductory readings in biology.* Boston: Little Brown.

Report of the Panel on Youth of the President's Science Advisory Committee. (1974). *Youth: Transition to adulthood.* Chicago: University of Chicago Press.

Resnick, L. B. (1987). *Education and learning to think.* Washington, DC: National Academy Press.

Resnick, L. B., & Wirt, J. G. (1996). *Linking school and work: Roles for standards and assessment.* San Francisco: Jossey-Bass.

Restivo, S. (1994). *Science, society, and values: Toward a sociology of objectivity.* Bethlehem, PA: Lehigh University Press.

Rubinstein, E. (1994). Science innovations on campus. *Science, 266* (5184), 843–893.

Rutter, M., Maughan, B., Mortimore, P., & Ouston, J. (1979). *Fifteen thousand hours.* Cambridge, MA: Harvard University Press.

SCANS (1991). *What work requires of schools.* Washington, DC: U.S. Department of Labor.

SCANS (1993). *Skills and tasks for jobs: A SCANS report for America 2000.* Washington, DC: U.S. Government Printing Office.

S. 2114, 101st Cong., 2d Sess. (1990). Washington, DC: U.S. Government Printing Office.

S. Rep. No. 3095, 101st Cong., 2d Sess. (1990). Washington, DC: U.S. Government Printing Office.

S. Rep. No. S.2 (An Act), 102d Cong., 2d Sess. (1992). Washington, DC: U.S. Government Printing Office.

Schmidt, W. H., McKnight, C. C., & Raizen, S. A. (1997). *A splintered vision: An investigation of U.S. science and mathematics education.* Boston, MA: Kluwer Academic Publishers.

Schrödinger, E. (1944). *What Is Life?* London, England: Cambridge University Press.

Scientific papers. (1991). *Science, 259* (5092), 180.

Scribner, R. A., & Chalk, R. A. (Eds.). (1977). *Adapting science to social needs: Knowledge, institutions, people into action.* American Association for the Advancement of Science.

Secretary's Commission on Achieving Necessary Skills. (1991). *What work requires of schools.* Washington, DC: U.S. Government Printing Office.

Secretary's Commission on Achieving Necessary Skills. (1992a). *Learning a living: A blueprint for high performance.* Washington, DC: U.S. Government Printing Office.

Secretary's Commission on Achieving Necessary Skills. (1992b). *Skills and tasks for jobs.* Washington, DC: U.S. Government Printing Office.

Secretary's Commission on Achieving Necessary Skills. (1993a). *Skills and tasks for jobs: A SCANS Report for America 2000.* Washington, DC: U.S. Government Printing Office.

Secretary's Commission on Achieving Necessary Skills. (1994). (Updated). *Skills and tasks for jobs.* Washington, DC: U.S. Government Printing Office.

Shadish, W. R., & Fuller, S. (Eds.). (1994). *The social psychology of science.* New York: Guilford Press.

Sherman, S. W. (1983). *Education for tomorrow's jobs.* Washington, DC: National Academy Press.

Shymansky, J. A., & Kyle, W. C., Jr. (1990). *Establishing a research agenda: The critical issues of science curriculum reform.* Iowa City, IA: The University of Iowa Science Education Center.

Silberman, C. E. (1970). *Crisis in the classroom: The remaking of American education.* New York: Random House.

Smithsonian Institution and National Academy of Sciences. (1996). *Resources for teaching elementary school science.* Washington, DC: National Academy Press.

Snow, C. P. (1959). *The two cultures and the scientific revolution.* New York: Cambridge University Press.

Spector, B.S. (1993). Order out of chaos: Restructuring schooling to reflect society's paradigm shift. *School Science and Mathematics, 93,* 9–19.

Spencer, H. (1859). *Education: Intellectual, moral, and physical.* New York: John B. Alden.

Spiegel-Rösing, I., & Price, D. (Eds.). (1977). *Science, technology, and society: A cross-disciplinary perspective.* Beverly Hills, CA: Sage.

Stoller, K. P. (1994). The use of animals in laboratory research: Debate presses forward. *The Scientist, 81,* 12.

Teich, A. H., Nelson, E. D., & McEnaney, C. (Eds.). (1994). *Science and technology policy yearbook, 1993.* Washington, DC: American Association for the Advancement of Science.

Textley, J., & Wild, A. (1996). *NSTA Pathways to the science standards: High school edition.* Arlington, VA: National Science Teachers Association.

Thayer, V. T. (1938). *Science in general education.* New York: Appleton-Century.

Toffler, A. (1970). *Future shock.* New York: Random House.

Trigg, R. (1985). *Understanding social science.* New York: Basil Blackwell.

UNESCO-NIER. (1996, November). Educational targets for the twenty-first century. *NIER Newsletter, 28:* 3.

U.S. Congress, 98th, 1st Session. (1983). *U.S. children and their families.* Report of Select Committee on Children, Youth, and Families, House of Representatives. Washington, DC: U.S. Government Printing Office.

U.S. Congress, 100th, 1st Session. (1987). *U.S. children and their families: Current conditions and recent trends.* House of Representatives. Washington, DC: U.S. Government Printing Office.

U.S. Congress, Office of Technology Assessment. (1991). *Adolescent health* (Vols. 1–3). Washington, DC: U.S. Government Printing Office.

U.S. Department of Education. (1991). *Youth indicators 1991: Trends in the well-being of American youth.* Washington, DC: U.S. Government Printing Office.

Urban, W. J. (1990). Social and institution analysis. *American Educational Research Journal, 27* (1), i–vi.

Veggeberg, S. (1993). Convergence of disciplines propels cognitive science. *The Scientist, 7,* 15.

Vignettes: retelling. (1993). *Science, 259* (5093), 390.

Wallace, B., & Srb, A. M. (1961). *Adaptation.* Englewood Cliffs, NJ: Prentice-Hall.

Ward, S. A., & Reed, L. J. (Eds.). (1983). *Knowledge, structure and use: Implications for synthesis and interpretation.* Philadelphia: Temple University Press.

Webster, A. (1991). *Science, technology, and society: New directions.* New Brunswick, NJ: Rutgers University Press.

Weinberg, A. M. (1972). Science and trans-science. *Minerva, 10* (2), 209–222.

Wetzel, J. R. (1989). *American youth: A statistical shapshot.* Washington, DC: William T. Grant Foundation.

Whitehead, A. N. (1972). *Science and the modern world.* New York: Harper and Row.

Wiesner, J. B. (1962). *Strengthening the behavioral sciences* [A report to the White House by the Life Sciences Panel, President's Science Advisory Committee]. Washington, DC: U.S. Government Printing Office.

Wilkinson, J. J. G. (1847). *Science for all: A lecture.* London: William Newberg.

Wirth, A. G. (1992). *Education and work for the year 2000: Choices we face.* San Francisco: Jossey-Bass.

Wolpert, L. (1993). *The unnatural nature of science.* Cambridge, MA: Harvard University Press.

Woolf, H. (Ed.). (1964). *Science as a cultural force.* Baltimore, MD: Johns Hopkins Press.

Woolgar, S. (1988). *Science: The very idea.* New York: Tavistock Publications.

Ziman, J. (1968). *Public knowledge: The social dimension of science.* Cambridge, UK: Cambridge University Press.

Ziman, J. (1990). *Teaching and learning about science and society.* New York: Cambridge University Press.

Index

About the Author

Paul DeHart Hurd received his EdD in science education from Stanford University in 1949. He holds honorary Doctor of Science degrees from the University of Northern Colorado, 1980; Ball State University, 1979; and Drake University, 1974. From 1929 to 1949 he taught high school and junior college science courses. At Stanford University he was in charge of teacher education programs in science. His research has focused on the history of science curriculum reform movements in the United States and foreign countries. He is the author of nine books and monographs on science education and some 200 published articles.